A Guide to a Healthier Planet, Volume 2

Erlijn van Genuchten

A Guide to a Healthier Planet, Volume 2

Scientific Insights and Actionable Steps
to Help Resolve Climate, Pollution
and Biodiversity Issues

 Springer

Erlijn van Genuchten
Sustainable Decisions
United Nations Economic Commission for
Europe Task Force on Digitalization in Energy
Tübingen, Germany

ISBN 978-3-031-60127-9   ISBN 978-3-031-60128-6   (eBook)
https://doi.org/10.1007/978-3-031-60128-6

This Springer imprint is published by the registered company Springer Nature Switzerland AG
The registered company address is: Gewerbestrasse 11, 6330 Cham, Switzerland

If disposing of this product, please recycle the paper.

For nature

Foreword

I am part of a generation that was growing up in an era of growth, optimism, and the rush to consume. A generation that only later in life realized that there is no "Planet B"… As a child, I experienced actual scarcity of food, clothing, and experiences. For example, my first trip abroad happened when I was 13, and not only was I excited to get on a big plane, but I felt like I wanted to do it over and over again. This was in the 1990s.

Decades later, I see things differently. I have experienced environmental degradation and what it means for health, for example living in a beautiful house with a beautiful view but with air full of smoke from hay being burned on the fields. I have worked on it, through my years at the United Nations Development Programme (UNDP) and now as Deputy Executive Secretary at the United Nations Economic Commission for Europe (UNECE), where we try to promote norms and standards for air quality, help countries prevent industrial pollution, and build cleaner vehicles. And, finally, I have heard about it from my children: much more sensitive and aware—a very different generation from mine.

So, reading *A Guide to a Healthier Planet*—Volume 2, I found that it is doing an incredibly important job: it communicates about complex issues in a way that is understandable, that resonates, and that enables action. And it does so on a solid scientific base. Why is this important?

First, because policy choices are too often informed by oversimplified notions about human and planetary health. These oversimplified notions, coming from the (social) media, may overshadow the importance of properly researched but hard-to-comprehend policy briefs prepared by professionals. Instead, a policymaker would do well by reading this Guide—putting things in a scientific perspective but without requiring rigorous study.

Second, because we need to be able to illustrate the complexity of the interdependencies underpinning the triple planetary crisis. The effects of nitrogen pollution on allergies, the effects of the changing climate on mountain ecosystems and ecosystems of pollinators, the unique role of whales in the global ecological balance all are important (and not at all obvious!) examples of such interdependencies. They point to a need for more subtle and more systemic solutions.

Third, because we need to be able to talk about solutions in this clear way too. Importantly, this Guide offers solutions at several levels:

(a) Personal action and responsibility (*yes, switching to rail travel makes a big difference!*)
(b) Industry innovation and new business models (*be it self-healing concrete or non-fossil-fuel based kerosene or artificial intelligence-powered waste sorting*)
(c) Future of policy and regulation (*which needs to reflect the links between climate and health; pollution and biodiversity; etc., and hence needs to be much less siloed*)

Dr. van Genuchten masterfully delivers on all three fronts. Her book would be a useful and inspiring read for a minister, an international activist, a CEO, or you and I, just to name a few. The videos in-built into the text provide for a useful mix of media, and the well-structured narrative helps navigate the reading, but also makes the book an actual guide.

So, with this Guide at our disposal, let's work together to keep our planet safe and beautiful—for ourselves, our children, and generations yet to come.

United Nations Economic Commission for Europe.[1]

Geneva, Switzerland   Dmitry Mariyasin

---

[1] The views described here are Dmitry Mariyasin's own and do not necessarily represent the opinion or endorsement of the UNECE.

Introduction

First, I would like to thank you! Thank you for taking the time to learn more about one of the most pressing topics in today's world: the health of our planet! It is an incredibly important topic because our beautiful planet Earth is what provides us with the resources to live. Without these resources, all other issues we deal with in daily life are irrelevant. This is supported by Maslow's hierarchy of needs (see Fig. 1), showing the different needs in life, with physiological needs at the base of the pyramid. The pyramid expresses that the lower needs must be satisfied before higher needs become relevant.

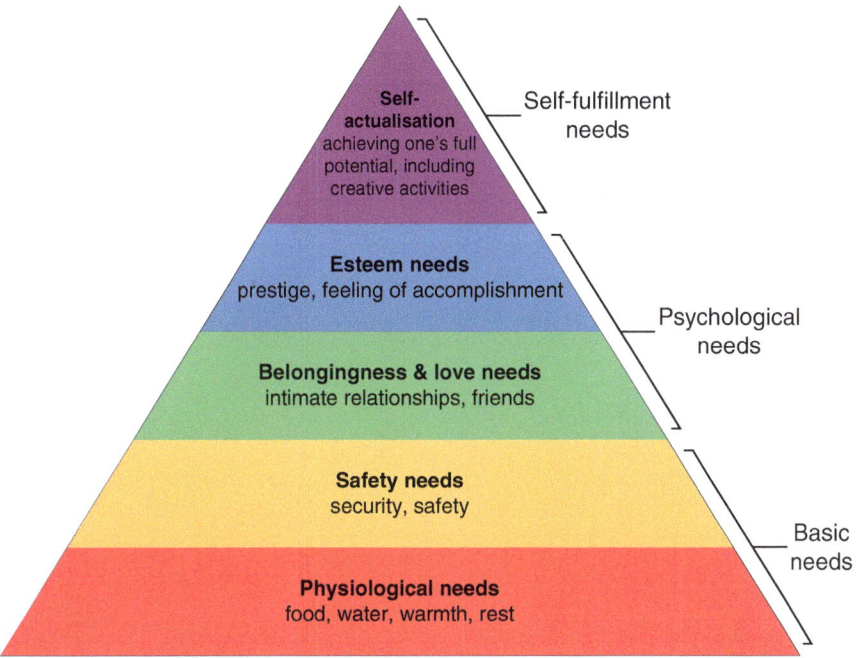

Fig. 1 Maslow's hierarchy of needs

And even though not everyone is aware of it or already experiencing it, some of these physiological needs are at risk. This is because we are currently facing a triple planetary crisis. Triple means that we have to deal with three environmental crises at the same time: climate change, environmental pollution, and biodiversity loss:

1. Climate change refers to changes in temperature and weather patterns caused by rising levels of greenhouse gases such as carbon dioxide (CO_2) and methane in the atmosphere.
2. Environmental pollution refers to physical, chemical, and biological contaminants that harm the earth or atmosphere to such an extent that normal processes are disrupted. An example of a physical contaminant is noise, a chemical contaminant is sunscreen, and a biological contaminant is a virus.
3. Biodiversity loss refers to different plant and animal species going extinct.

A broad range of issues and the far-reaching consequences caused by the triple planetary crisis have been discussed in Volume 1 of *A Guide to a Healthier Planet*. For example, Volume 1 explains how climate change impacts our wine, how heavy metal pollution can contribute to Parkinson's disease, and how soil biodiversity impacts the health of our planet. Also, it for example explains how we can mitigate climate change by reducing methane emissions, how we can remove pollutants from water and soil, and how biodiversity can be protected by more sustainable agriculture.

While it is good news that solutions exist for each of the consequences discussed in Volume 1, it is important to realize that many more issues are caused by these environmental crises. And that many more solutions are available. Many of these solutions can be put into practice by us as individuals in daily life. And even when these actions may seem small, they are extremely important. In the video in Fig. 2, I explain why every single action counts.

To further increase our understanding of the broad range of issues and solutions, Volume 2 of *A Guide to a Healthier Planet* covers further issues and solutions related to climate change, pollution, and biodiversity loss. The structure of the book is the same, with each part focusing on one of the three crises. Again, each part consists of four chapters addressing examples of current and future issues and two chapters addressing examples of how we can resolve these issues and take action towards a healthier planet.

In the climate change part (see Table 1), we first look at how we can tell that cyclone intensity is impacted by climate change (Chap. 1). After that, we look at

Fig. 2

Table 1 Overview of Part I

Consequences	Chapter 1: Impact of Climate Change on Cyclone Intensity
	Chapter 2: Impact of Climate Change on Mountains
	Chapter 3: Impact of Climate Change on Animals
	Chapter 4: Impact of Climate Change on Pregnant Women and Birth Defects
Solutions	Chapter 5: Controlling CO_2 Levels in the Building Sector
	Chapter 6: Controlling CO_2 Levels in the Aviation Sector

Table 2 Overview of Part II

Consequences	Chapter 7: Impact of Different Types of Pollution on Pollinators
	Chapter 8: Impact of Sunscreen Pollution on Marine Environments
	Chapter 9: Impact of Nitrogen Pollution on Pollen Allergies
	Chapter 10: Impact of Plastic Pollution on Spreading Viruses
Solutions	Chapter 11: Removing Plastic Waste from the Environment
	Chapter 12: Removing Waste from Landfills

Table 3 Overview of Part III

Consequences	Chapter 13: Impact of Amazon Deforestation on Biodiversity
	Chapter 14: Impact of Exotic Animal on Ecosystems
	Chapter 15: Impact of Global Warming on Ocean Biodiversity
	Chapter 16: Impact of Whales on Our World
Solutions	Chapter 17: Protecting Natural Ecosystems
	Chapter 18: Sustainable Fishing

consequences that we are already clearly noticing in daily life: how climate change affects mountains (Chap. 2) and animals (Chap. 3). Then we look at consequences that may not be so obvious: how pregnant women and unborn babies are affected by climate change (Chap. 4). In the last two chapters of this part, we look at solutions: at how CO_2 emissions can be reduced in the building (Chap. 5) and aviation sectors (Chap. 6).

In the pollution part (see Table 2), we first look at how different types of pollution affect pollinators (Chap. 7) and how sunscreen pollution affects marine environments (Chap. 8). After that, we look at how pollution also directly impacts us: how nitrogen affects pollen allergies (Chap. 9) and plastic pollution can spread viruses (Chap. 10). In the last two chapters of this part, we look at solutions: how plastic pollution can be removed from the environment (Chap. 11) and how waste from landfills can be removed (Chap. 12).

In the biodiversity part (see Table 3), we first look at the consequences of cutting down trees in the Amazon rainforest on plants and animals (Chap. 13) and how exotic animals harm ecosystems (Chap. 14). After that, we look at the consequences

Fig. 3 Sustainable living is not only easier when understanding the importance and psychology behind it but also rewarding

of global warming on ocean biodiversity (Chap. 15) and how whales have a huge impact on our world (Chap. 16). In the last two chapters of this part, we look at solutions: how natural ecosystems can be protected (Chap. 17) and how the fishing industry can be made more sustainable (Chap. 18).

Finally, in the conclusion is discussed why today's 'grow now, clean up later' mentality has devastating consequences and why adopting an environmental protection mentality is important. Interestingly, understanding the importance of sustainable living and the psychology behind it not only makes it easier but also more rewarding to contribute to a healthier planet (see Fig. 3)!

As in Volume 1, each chapter is based on one or more recent scientific publications and makes scientific insights from these publications available in easy-to-understand language. In addition, in each chapter, ideas are added about what you and I can do in daily life to make a positive difference. This allows each and every one of us to take the first or next step straight away.

Figure Credits

Acknowledgement

I would like to thank the many wonderful people in my life who have supported me in my sustainability journey from the beginning, which eventually allowed me to write this book. In particular, I would like to thank my wonderful partner, family, friends, and coaches Ken Porter (dec.), Gil McIff, Marci Meyers, Jed Pfaff, Alexis Bird, and Karina von Keitz, as they have been there for me in good and hard times for a long time. I am also very grateful for the thoughts and input Sheryl Larson, Rameen Ashraf Ali, Cristina Solis, Kanchana Peeris, Kristina Zuna, and Amy Meleca contributed to the content of this book. Also, I would like to thank the many scientists who are doing an awesome job in providing insights that help us reach a more sustainable future and Dmitry Mariyasin for writing an excellent foreword. Finally, I would like to thank nature for providing me the necessary resources to be able to live on this gorgeous planet Earth!

Contents

Part IV   Conclusion

Part I
Climate Change

With the climate crisis progressing, the disastrous consequences are becoming more obvious in daily life and are present in the news more frequently. For example, extreme weather events such as storms and floods made many headlines in 2023, as ten countries on different continents experienced severe flooding in less than two weeks.

As further extreme weather events and resulting catastrophes are likely to occur in the near future, they are also a call for immediate action toward mitigating climate change. To trigger climate action, several climate change meetings, such as the United Nations' Climate Conference series COP, have led to agreements that define climate goals. One of these agreements is the Paris Climate Agreement, which aims at limiting global warming to less than 2 °C (3.6 °F), ideally 1.5 °C (2.7 °F). Also, the European Union has committed itself to achieving net-zero greenhouse gas emissions

Fig. 1 Net-zero means that as much greenhouse gas emissions are extracted from (*left part of the scale*) as sent into (*right part of the scale*) the atmosphere

Fig. 2 As climate policies are not bringing about required action to mitigate climate change, actions taken by individuals like you and I are critical

by 2050. A net-zero emission goal means that they strive to balance out the amount of greenhouse gases sent into and extracted from the atmosphere (see Fig. 1).

While the awareness and concern about climate change are spreading, a large difference exists between commitments and actions taken. This is called the knowledge-action gap. In fact, this knowledge-action gap has not been improved by climate policies at a national level across the EU since the Paris Agreement was adopted in 2015. In some European Union member states, this knowledge-action gap is even widening!

So, while the call to take action is an old one, it is more relevant than ever. And as governments are not enabling the required speed of change, we as individuals need to take our future into our own hands (see Fig. 2).

Credit

This Chapter Is Based On:

Nayna Schwerdtle, P., et al. & Jungmann, M. (2023). Interlinkages between climate change impacts, public attitudes, and climate action—Exploring trends before and after the Paris Agreement in the EU. *Sustainability*, *15*(9), 7542.

Figure Credits

Chapter 1
How Climate Change Impacts Cyclone Intensity

Abstract In recent years, more and more countries across the globe experienced severe flooding due to cyclones, typhoons, or hurricanes. While such storms naturally impacted the world throughout history, their increased intensity is caused by climate change. This trend can be shown using different methods. These methods involve indices, historical datasets, observations, and numerical modeling. To be able to draw high-quality conclusions about this trend, enough high-quality data needs to be available. Also, statistical tests must be performed to ensure that changes are not caused by chance. The resulting insights help us understand the effects of climate change on future cyclones and their consequences.

Keywords Science · Science communication · Climate change · Climate change consequences · Global warming · Tropical cyclone · Typhoon · Hurricane · Extreme weather events · Intensity change

In 2023, ten countries on different continents experienced severe flooding in less than 2 weeks. Some of these floods were caused by storm Daniel, which severely affected several countries including Greece and Libya (see Fig. 1.1). Other floods were caused by multiple typhoons in Asia, also affecting several countries including Taiwan, China, and Hong Kong. And also North and South America were affected. The consequences involved streets that turned into deadly rivers, villages being submerged, collapsed dams, flooded metro stations, and homes without power. Just to name a few.

The storms causing these devastating effects were very strong, destructive storms with winds rotating inward (see Fig. 1.2). Wind speeds are 119 km/h (74 mph) or more, with the highest speeds and heaviest rainfall near the center, or eye, of the typhoon. Depending on where they form, these storms are called typhoons, cyclones, or hurricanes: typhoons form over the Northwest Pacific, cyclones over the South Pacific and Indian Ocean, and hurricanes over the North Atlantic, central North

---

Credit: This chapter is based on two scientific articles by Liguang Wu and Nick Marriner and their colleagues. (Full citations are available at the end of the chapter)

Fig. 1.1

Fig. 1.2 Cyclones are characterized by strong, destructive winds

Pacific, and eastern North Pacific. In this chapter, I use cyclone as general term for these storms, focusing on those that start around the equator in tropical areas as indicated by the dark band in Fig. 1.3.

While cyclones and resulting floods have impacted the world throughout history, an important question is whether this series of extreme weather events is a coincidence or whether it is caused by anthropogenic climate change. Anthropogenic climate change means changed climatic conditions due to human activities as opposed to natural changes. As cyclones occur naturally and each cyclone is different, it can be easy to overlook the effects of climate change. But understanding these effects is very important because with increasing rainfall, intensity, and number of cyclones globally, the damage and number of disasters and deaths will increase.

To be able to understand the effects of climate change on cyclones and future consequences, it is important to know which factors influence the intensity and structure of cyclones. Here, it makes a difference whether individual, large-scale, or global cyclones are studied as factors can differ on different scales. For example, Sahara sand dust can reduce cyclones over the Atlantic, and may be able to prevent large-scale cyclone formation but cannot stop cyclone development on a global scale. This is because the effect of dry and warm Sahara air with strong vertical winds can impact cyclones only locally.

This is how we can tell climate change impacts the intensity of large-scale tropical cyclones:

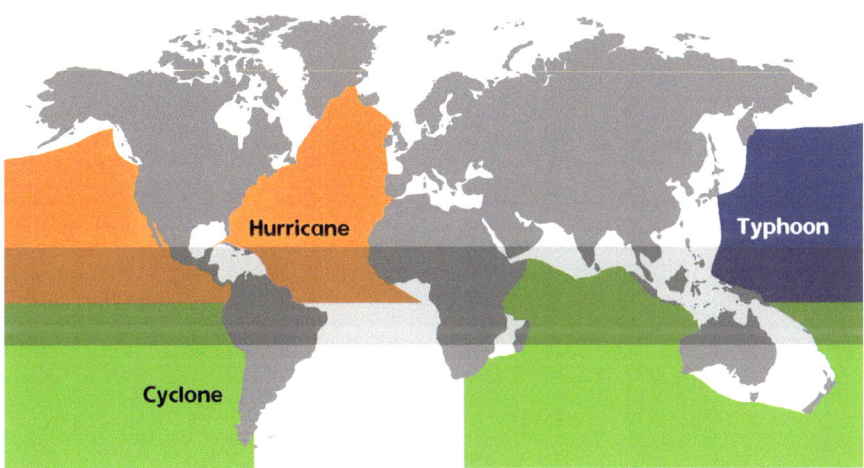

Fig. 1.3 The difference between hurricanes, typhoons, and cyclones is where they form, roughly indicated by different colors: *orange* indicates where hurricanes, *blue* where typhoons, and *green* where cyclones are formed

1.1 Indices

The first way we can tell that climate change impacts large-scale tropical cyclones is by using indices. An index is a measure of certain properties.

The first type of index used for cyclones measures the maximum wind speed. Wind speed is measured on a five-point scale, with higher values meaning intenser cyclones. With this index, the average maximum wind speed over all cyclones in a year can be calculated. But also how often cyclones with a certain level occur in a year. Alternatively, the proportion of cyclones at a certain level can be calculated. The advantage of using a proportion instead of an absolute value is that the influence of changes in the total number of cyclones is removed. For example, when 5 cyclones out of 10 in a year are of level 4, and 10 years later it was 5 out of 5, the absolute number remains the same, but the proportion changes from 50% to 100%.

The second type of index used for cyclones measures the combination of frequency, duration, and intensity. One of the indexes that uses these three parameters is the Accumulated Cyclone Energy index. This index was used to visualize cyclone intensity for the North Atlantic area in this graph in Fig. 1.4. Higher bars mean higher overall intensity of hurricanes in a hurricane season.

The third type of index used for cyclones measures the maximum potential intensity of a cyclone. To calculate the maximum potential intensity, relationships between environmental parameters and cyclone intensity are taken into account. This is helpful because apart from internal factors, also several environmental parameters impact tropical cyclone intensity. For example, environmental factors that slow cyclones down are:

Accumulated cyclone energy of North Atlantic hurricanes

Accumulated cyclone energy (ACE) is an index used to measure the activity of a cyclone/hurricane season. It combines the number of hurricane systems, how long they existed and how intense they became. It is calculated by squaring the maximum sustained surface wind in the system every six hours that the cyclone is a Named Storm and summing it up for the season.

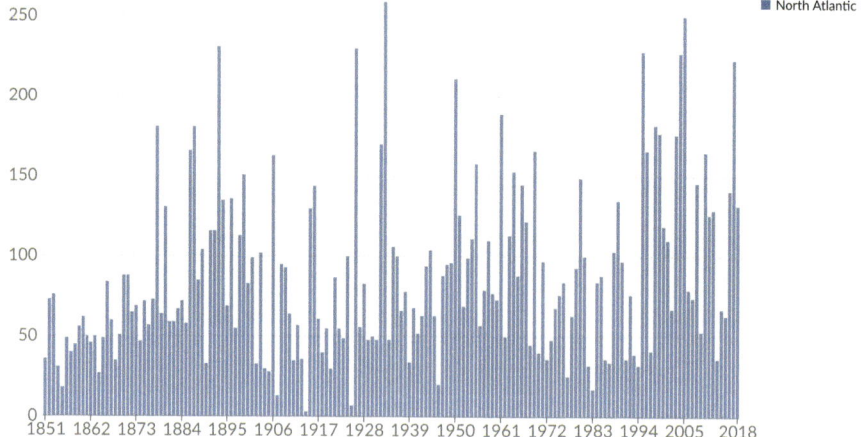

Fig. 1.4 The Accumulated Cyclone Energy index for the North Atlantic area between 1851 and 2018

- vertical winds hitting the horizontal winds of the cyclone
- dry air that is sucked into the storm
- air that is cooled by the sea or ocean

By measuring environmental circumstances today, this index allows us to predict future cyclone intensities and tell how intensities will change over time.

1.2 Historical Datasets

The second way we can tell that climate change impacts large-scale tropical cyclones is by using historical datasets. These datasets rarely contain measures of wind speed or pressure high up in the air, as these could not be measured before airplanes existed. That is why measures from these datasets contain observations from ships at sea and coastal weather stations.

After World War II, airplanes were used to measure cyclone intensities. But with limitations, as aircraft measurements contain several biases, for example depending on the instrumentation used for measuring, flight wind, and flight height. Also, limitations are caused by evolving technology and developments in how tropical cyclone intensity is calculated, as these changes make comparing calculations from different years challenging.

Fig. 1.5 Cyclones are measured using satellites

Because of these limitations, cyclone intensities and locations have been measured using satellites since 1979. Satellite measures are more consistent and make
more recent datasets relatively reliable. That is why data from satellites can be used
to create computational models that take different environmental factors into
account (see Fig. 1.5). A computational model uses computer programs to simulate
and study complex systems using different types of calculations. The accuracy of
these models can be measured by comparing the output of models with real data.
Accurate models allow scientists to study the impact of different environmental
parameters on cyclone frequency and intensity.

1.3   Observations

The third way we can tell that climate change impacts large-scale tropical cyclones
is by using observations. Here, observational evidence should have high quality. If
not, conclusions cannot be of high quality either. Also, data must be sufficient as
natural variability exists. If the time frame is too short, incorrect conclusions can
be drawn.

When high-quality and enough data is available, it is important to use statistical
analyses to make sure that changes are large enough and not caused by chance. For
example, as data from the last few years is missing in the graph in Fig. 1.6, it could

Power Dissipation Index (PDI) of North Atlantic cyclones

The Power Dissipation Index (PDI) measures the activity of cyclones by accounting for cyclone strength, duration, and frequency. The lines have been smoothed using a five-year weighted average, plotted at the middle year.

Data source: HURDAT (Hurricane Database) of the National Oceanic & Atmospheric Administration (NOAA)
OurWorldInData.org/natural-disasters | CC BY

Fig. 1.6 As data from the last few years is missing, it could be falsely concluded that cyclones are becoming less intense

be falsely concluded that cyclones are becoming less intense. And whether there is a difference over time needs to be confirmed using statistical tests.

For example, historical data was used to tell whether climate change is impacting cyclones over the Mascarene Islands (see Fig. 1.7). These islands are located in the Indian Ocean, about 850 km (528 mi) east of Madagascar. For this area, a 300-year historical record with high-quality and consistent data exists about tropical cyclones. This record shows a fourfold increase in cyclones since 1940: before 1940, cyclones occurred on average about 3 times per 20 years, whereas after that, they occurred about 14 times per 20 years. This is a statistically significant difference, which means that this is a trend and not caused by mere chance.

1.4 Numerical Modeling

The fourth way we can tell that climate change impacts large-scale tropical cyclones is by using numerical modeling. Numerical modeling involves computers running many mathematical calculations to find an approximate solution to a physical problem. The difference with computational models is that numerical models focus on relations, variables, and magnitudes, instead of how something works. Numerical modeling can be done in several ways:

Fig. 1.7 Tropical cyclone over Mascarene Islands, with La Reunion still visible

The first way to use numerical modeling is by combining models on a regional scale, for example per 18 km (11.2 mi) with models on a global scale. A model is in this context a mathematical representation of the world. The regional models simulate a cyclone while using environmental parameters from the global models. This makes it for example possible to simulate cyclones with high CO_2 levels and cyclones with low CO_2 levels in the atmosphere. When comparing these simulations, it is possible to tell that global warming caused by high CO_2 levels leads to more frequent and more intense cyclones.

The second way to use numerical modeling is by combining models that predict how cyclones form, move, and how intense they are. Within these models, further parameters are included. For example, the models that predict cyclone intensity take temperature, ocean changes, and vertical wind into account. The advantage of combining models is that the relative importance of different parameters can be measured. When combining these models, it also becomes clear that cyclone intensity will increase over time.

The third way to use numerical modeling is by using global models of the atmosphere. This means that certain factors are not taken into account, such as how the ocean influences cyclones. These models also predict increased frequency and intensity of cyclones. And as computers are becoming more and more powerful, it is now also possible to combine different global models for more accuracy.

1.5   Conclusion

So, extreme weather events do not only seem to becoming more common in the news, scientific results confirm that cyclone intensity is increasing due to climate change. These scientific results are based on different measures. These measures include indices that capture values for different properties such as wind speed, and relationships between environmental parameters and cyclone intensity. Numerical modeling can be used to combine relations, variables, and magnitudes from different models, to gain a better understanding of the effects of climate change.

Also, historical data sets can be used as they allow us to compare today's cyclones with cyclones in the past. These comparisons can also help uncover trends about how cyclones are developing over time. Here, historical data must be high-quality and contain enough data to be able to draw high-quality conclusions. These conclusions help to further understand the effects of climate change on future cyclones and their consequences.

1.6   How We Can Take Action

As climate change has a large impact on cyclones, contributing to limiting climate change is extremely important. Here are practical ideas of what you and I can do to slow down climate change and minimize the disastrous effects of cyclones:

- Reducing CO_2 emissions as much as possible, for example by buying fewer clothes, traveling by public transport as much as possible, and eating vegan or vegetarian meals
- Using building materials for homes that are resistant to strong winds
- Taking warning signals seriously, so that you can bring yourself and others into safety
- Preventing littering, as waste can become a projectile during a storm
- Supporting cyclone victims, both on a practical and mental level (further reading: Chap. 3 of A Guide to a Healthier Planet Volume 1: "How Climate Change Impacts Mental Health")

Credit

This Chapter Is Based On:

Marriner, N., Kaniewski, D., Garnier, E., Pourkerman, M., Giaime, M., Vacchi, M., & Morhange, C. (2022). Has the Anthropocene affected the frequency and intensity of tropical cyclones? Evidence from Mascarene Islands historical records (southwestern Indian Ocean). *Global and Planetary Change, 217*, 103,933.

Wu, L., Zhao, H., Wang, C., Cao, J., & Liang, J. (2022). Understanding of the effect of climate change on tropical cyclone intensity: a review. *Advances in Atmospheric Sciences, 39*(2), 205–221.

Figure Credits

Chapter 2
How Climate Change Impacts Mountains

Abstract Climate change has a significant impact on landscapes, including mountains. As global temperatures rise, snow, glaciers, and permafrost melt, which further disrupts the regional heat balance. This disruption increases the number of geohazards and risks, such as landslides, rock slope failures, and falling rocks. Also, mountain ecosystems and services change, for example affecting plant growth, water quality, and food web structures. This in turn strongly impacts mountain communities and infrastructure, as they rely on mountains for water, food, agriculture, and tourism. This shows that higher global temperatures have far-reaching consequences in mountain areas.

Keywords Science · Science communication · Climate change · Climate change consequences · Global warming · Mountains · Mountain ecosystem · Glaciers · Permafrost · Geohazards · Ecosystem services · CO_2 · Communities · Landscape · Food web · Elevation

While cyclones are one way for us to notice that climate change is progressing, the consequences are also noticeable around us independent of extreme weather events. This is possible because climate change has many different consequences. For example, climate change also has a large impact on landscapes and their ecosystems. An ecosystem is a community of organisms that interact with each other and with the environment. One of these ecosystems is the marine environment: 71% of our planet's surface is covered by oceans. Another important ecosystem is mountains: of the 29% of land surface on our planet, about 15% is higher than 1000 m above sea level (3280 ft or 0.6 mi)!

The impact on landscapes and their ecosystems can be from a small to large scale. An example of the impact on a small scale is provided in Chap. 2 of A Guide to a Healthier Planet Volume 1 ("How Climate Change Impacts Our Wine"), which explains how vineyards and grape growth are affected.

Credit: This chapter is based on the scientific article "Scientists' warning of the impacts of climate change on mountains" by Jasper Knight. (Full citation is available at the end of the chapter)

Fig. 2.1 Mountains are also highly impacted by global warming

An example of the impact on a large scale can be seen in mountains. Mountain ecosystems are currently getting out of balance due to anthropogenic climate change, as snow and ice are melting. Snow and ice play an important role in keeping the regional heat balance. This heat balance is kept when light-toned snow and ice surfaces reflect sunlight into outer space and the surface stays cool.

As global temperatures are rising due to climate change, snow and ice are melting, which disturbs this heat balance. As a consequence, other aspects of mountain ecosystems are changing too, which causes further aspects of the mountain ecosystem to get out of balance. These other aspects include wind direction and humidity. These changed environmental circumstances all impact mountains significantly with far-reaching consequences. What the consequences are and to what extent, differ between mountains, as mountains are complex systems consisting of many parts, and not all are equally sensitive to climate change (see Fig. 2.1). This is how mountains and their ecosystems are impacted:

2.1   Glaciers

The first aspect of mountains that is highly impacted by anthropogenic climate change is glaciers. A glacier is a slowly moving river of ice (see Fig. 2.2).

Glaciers are impacted by climate change because rising temperatures cause glaciers to shrink, become thinner, and move differently (see Fig. 2.3). How exactly and to what extent they change depends on many factors, including the glaciers'

Fig. 2.2 A glacier

Fig. 2.3 *Left half*: comparison of Mount Stanley in the Democratic Republic of Congo in 1906 (*top*) and 2022 (*bottom*); *right half*: comparison of West Stanley Glacier in Canada 4910 m (16,109 ft) above sea level in 2012 (*top*) and 2022 (*bottom*); clearly showing less snow and ice

location, elevation, climate, and size. For example, mountain glaciers are relatively sensitive to higher temperatures when they are small in size and steep. And glaciers shrink less quickly on higher mountains.

When predicting how glaciers will change or whether they will melt completely, not only temperature should be taken into account. Also for example the amount, size, and distribution of debris, including small rocks, that end up on the glacier influence changes. This is because dark-toned surfaces, such as rocks, absorb sunlight, which causes the surface to warm up. These dark surfaces cause snow and ice to melt quicker, in turn causing areas with melting snow and ice to heat up about three times quicker than areas without ice.

2.2 Permafrost

The second aspect of mountains that is highly impacted by anthropogenic climate change is permafrost. Permafrost is ground that is frozen for at least two years in a row.

Permafrost is impacted by climate change because rising temperatures cause the thickness of the seasonally frozen layer above permafrost to change and subsurface temperatures to rise. For example, the permafrost temperature in the Tibetan Plateau with an average height of 4500 m (14,800 ft) is expected to increase by about 2.6 °C (4.7 °F). As a consequence, the thickness of the permafrost earth layer will decrease by 2 to 4 m (6.6 to 13.1 ft) by 2100. Also, the height above sea level at which permafrost starts increases.

When predicting how permafrost will develop and whether it will melt, not only temperature should be taken into account. Also location specific slope properties, such as the amount of moisture and debris size impact permafrost and rock slopes. For example, how often the Mont Blanc mountain in France freezes is expected to be up to half as often in 2100 depending on altitude. This will impact the stability of the surface and how the mountain's slopes will change.

2.3 Geohazards and Risks

The third aspect of mountains that is highly impacted by anthropogenic climate change is mountain geohazards and risks. Mountain geohazards and risks are common in mountains because of earthquakes, volcano eruptions, steep slopes, a harsh climate, and the presence of snow and ice.

While these hazards and risks are unrelated to climate change, additional hazards and risks are caused by melting glaciers, permafrost, and snow as this makes land unstable and able to move. Land movements can result in landslides, rock slope failures and falling rocks, flowing debris, and fans (see Fig. 2.4) and terraces with all kinds of debris and rocks.

Also, more meltwater can cause floods. This is for example possible when the dam from a glacial lake bursts. These floods cause the earth to be full of water, which in turn can also lead to falling rocks, landslides, and debris or mud flows.

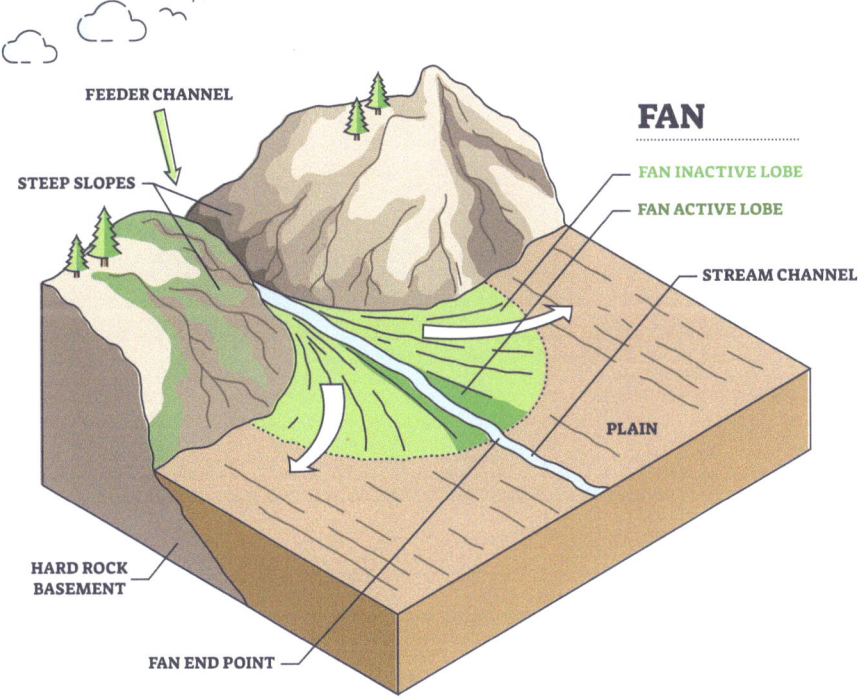

Fig. 2.4   A fan

2.3.1   Ecosystems and Services

The fourth aspect of mountains that is highly impacted by anthropogenic climate change is mountain ecosystems and services. Ecosystem services are the many benefits nature and healthy ecosystems provide us.

Ecosystems are impacted by climate change because they are very sensitive to direct changes, including higher temperatures and changes in rain and snowfall, and indirect changes through the impact on soil. These environmental changes can cause far-reaching consequences. At first, they for example allow plants to grow more days in a year, cause longer summers and fewer days with frost, and affect water quality. This in turn can for example allow certain plants to start growing higher up the mountain (see Fig. 2.5). This in turn can result in food web changes. A food web consists of all the food chains in the mountain ecosystem.

As a consequence, *services* are impacted as well. For example, as trees store a lot of carbon, changes in plant growth can influence how much carbon can be stored. Trees extract this carbon from CO_2 in the atmosphere. Also, with ecosystems changing, how easily diseases can spread changes as well.

Fig. 2.5 Anthropogenic climate changes causes changes in mountain vegetation

2.4   Communities and Infrastructure

The fifth aspect of mountains that is highly impacted by anthropogenic climate change is mountain communities and infrastructure. This is possible because people rely on mountains for example for water and food, as a place to live, for agriculture, and tourism.

Communities in mountain regions are impacted by climate change mostly because of differences in how much water is provided by glaciers and rivers. They need this water to for example drink, generate electricity, and grow food. With increasing temperatures, the amount of water that comes down is more variable, and with time, less water will be available.

For example, the Po River, the largest river in Italy, is drying up because not enough water is coming down from the Alps. This water is lacking because less snow is left on these mountains due to higher temperatures. Also, rain is falling within a shorter amount of time. In the video in Fig. 2.6, the consequences for people on agriculture, electricity production, and availability of drinking water living near the Po River are explained.

Infrastructure in mountain regions is also impacted by climate change because for example more water coming down can mean that ground moves and infrastructure gets damaged. For example, bridges can be affected by landslides (further reading: Chap. 4 of A Guide to a Healthier Planet Volume 1: "How Climate Change Impacts the Safety of Bridges").

Fig. 2.6

2.5   Conclusion

So, the impact of climate change on mountains and mountain ecosystems is diverse and far-reaching. This is because glaciers and permafrost are melting, causing the landscape to change. This changing landscape can cause additional hazards and risks in mountainous areas when for example rock slopes start to fail, rocks start to fall, too much meltwater causes floods, or lack of snow causes rivers to dry up. This in turn affects the mountain ecosystem and the services it provides, such as changed water quality and how likely diseases are spread. This in turn can affect mountain communities and infrastructure, for example when too much or too little water is available for drinking, agriculture, or power generation. Unfortunately, these consequences can become more severe when climate change progresses.

2.6   How We Can Take Action

As climate change has far-reaching consequences on mountains and mountain ecosystems, and eventually puts us at risk as well, contributing to limiting climate change and caring well for mountain ecosystems is critical. Here are practical ideas of what you and I can do to slow down climate change:

- Traveling to and from a mountainous region using environmentally friendly transport, such as traveling by train instead of flying
- Visiting mountains by foot instead of motorized vehicles
- Choosing sustainable accommodations that are resource-efficient

Here are practical ideas of what you and I can do to care well for mountain ecosystems:

- Preventing leaving any marks when visiting mountains, such as litter and food waste
- Respecting wildlife and not disturbing them
- Staying on marked trails
- Being careful to minimize campfire damage to the environment
- Keeping water clean
- Using as little water as possible

Credit

This Chapter Is Based On:

Knight J. (2022). Scientists' warning of the impacts of climate change on mountains. *PeerJ*, *10*, e14253.

Figure Credits

Chapter 3
How Climate Change Impacts Animals

Abstract Climate change is causing global temperatures to increase and regional and seasonal changes to become more common. These changing climatic conditions can impact animals in various ways. For example, these changes can influence the number of female sea turtles being born compared to the number of male sea turtles. And summer weather conditions influence how likely honey bees survive the winter. Also, climate change impacts hibernating animals in various ways, including their phenology, condition, reproduction, and survival. Depending on the species, age, and other factors, climate change can have both positive and negative impacts on hibernating animals.

Keywords Science · Science communication · Climate change · Climate change consequences · Global warming · Sex ratio · Sea turtles · Honey bees · Hibernation · Green Sea turtle · Hatching success · Incubation temperature · Marine turtles · Precipitation · Water availability · Nutrition · Pollinator · Foraging · Drought · Body mass · Conservation · Elevation · Evolution · Growth · Life history · Phenology · Reproduction · Survival

With climate change progressing, not only global temperatures are increasing, also regional and seasonal changes are becoming more common. While some are directly obvious, for example when wine is being affected by different climatic circumstances (further reading: Chap. 2 of A Guide to a Healthier Planet Volume 1: "How Climate Change Impacts Our Wine"), other consequences may not be so obvious. For example, temperature and seasonal changes can impact animals in different ways. Here are examples of how climate change affects animals:

Credit: This chapter is based on four scientific articles by Jacques-Olivier Laloë, Martina Calovi, Erin Wilson Rankin, and Caitlin P. Wells and their colleagues. (Full citations are available at the end of the chapter)

3.1 Impact on the Number of Female Sea Turtles

The first example of how climate change is affecting animals is by impacting the number of female sea turtles. It may be surprising, but environmental conditions play a crucial role in sea turtles' lives, especially while young turtles are still in their eggs. This is because sea turtles lay their eggs on beaches, in the sand (see Fig. 3.1). Some environmental factors such as global warming, heat this sand and other environmental factors such as shadow and rain, especially heavy rain, cool this sand down. When the sand cools down, the eggs cool down too.

The temperature of the eggs is important because what sex a baby turtle becomes depends on the temperature in the nest. The sex of a human child is determined when the sperm cell of the dad reaches the egg of the mum. But with turtles, this works differently: the sex of a baby turtle is determined in the incubation time, which is the 2 months in which the turtle is in its egg. To be more precise: the sex is determined during the middle third of these 2 months. This means that in the first few weeks, the turtle doesn't have a sex yet. Depending on the temperature in the nest during that time, more females are born when the temperature is higher, and more males are born when the temperature is lower. Already a small temperature difference can make a big difference: with a nest temperature of 30.8 °C (87.4 °F), almost all turtles are female. With a 1.1 °C (0.9 °F) lower temperature of 29.7 °C (85.5 °F), only 88 out of 100 are female. This can put the species at risk when due to climate change too few or hardly any male turtles are born.

Fig. 3.1 Sea turtle eggs

3.2   Impact on Honey Bees' Survival in Winter

The second example of how climate change is affecting animals is by impacting how likely honey bees survive in winter. Honey bees stay alive during the winter months because they rely on the nectar and pollen stores in their honeycomb. Whether they were able to fill their stores depends on the weather conditions in summer.

One aspect of summer weather conditions that is important for bees to survive in winter is the temperature. The temperature in summer is important for bees, as it influences how many flowers grow during the plant growing season. Both with too hot and too cold summers, there are fewer flowers from which they can collect pollen for proteins and nectar for carbohydrates. With fewer flowers due to for example droughts caused by climate change, they can collect less nectar. As they store this nectar as food for the winter, too hot and too cold summers mean honey bees may not be able to store enough food.

Also, the length of a hot summer affects whether honey bees survive the winter. Longer hot summers can help varroa mites to thrive (see Fig. 3.2). *Varroa mites* are parasites that live on bees. As they affect the bees' health by weakening them, the bees' chances for survival in winter become smaller.

Another aspect of summer weather conditions that is important for bees to survive in winter is the amount of rain. The amount of rain is important because when the summer is too dry, less water leads to lower protein quality of pollen, and to lower nectar quality and quantity. This means their food is less nutritious. When bees don't grow enough by the fall, they may not survive the winter. Also, too little or too much rain causes fewer or smaller flowers to grow. In this case, bees also have difficulties to gain enough weight. I assume this is because they have to fly further to gather the same amount of food.

Fig. 3.2 Bee larva with varroa mites

3.3 Impact on Hibernation

The third example of how climate change affects animals is by impacting hibernation. Hibernation is also called winter sleep, although hibernating animals don't
sleep. The similarity between hibernation and sleep is that bodily processes are
turned down to save energy; some main differences are that:

- the goal of sleep is resting whereas the goal of hibernation is surviving periods
  of for example food scarcity,
- hibernation lasts a lot longer than sleep, and
- the processes to keep the body alive are at a much lower level with hibernation
  than with sleep.

For example, groundhogs (see Fig. 3.3) start hibernating in late fall for about 3
months. During that time, their heart rate goes from about 80 beats per minute down
to 5 per minute. The number of breaths per minute goes down from 16 to 2 and their
body temperature from 37 °C (99 °F) to 3 °C (37 °F).

The changing climatic circumstances can impact hibernating animals in many
different ways, sometimes positive, sometimes negative, and sometimes it doesn't
seem to make a difference. This is how:

Fig. 3.3 Groundhog (*Marmota monax*)

3.3.1 Phenology

The first way changing climatic circumstances can impact hibernating animals is by affecting their phenology. Phenology involves their natural cycles, including their hibernation cycle. In general, warmer temperatures can lead to waking up earlier from hibernation and shorter hibernation periods for some animals but doesn't make a difference for other animals. On rare occasions, animals hibernate longer.

For example, rodents hibernate shorter with higher temperatures in spring. This means they wake up earlier in the year. For yellow-bellied marmots (see Fig. 3.4), this has positive consequences because they can grow longer, have a better body condition, and are more likely to survive.

The same is true for black bears (see Fig. 3.5) and brown bears (see Fig. 3.6): they tend to hibernate shorter with higher temperatures. Already with a 1 °C (1.8 °F) increase in the minimum winter temperature, black bears' hibernation period reduces by 6 days.

When temperatures change due to global warming and the hibernation period is shorter, waking up too early can cause a mismatch with other parts of the ecosystem. For example, when high-quality food is only available for a limited amount of time, this food may not be there yet. And even when enough food is available, it can mean that young animals have to care for themselves from an earlier age.

Fig. 3.4 A yellow-bellied marmot (*Marmota flaviventer*)

Fig. 3.5 Black bears (*Ursus americanus*)

Fig. 3.6 Brown bears (*Ursus arctos*)

3.3.2 Condition

The second way changing climatic circumstances can impact hibernating animals is by affecting their condition. For example, female arctic ground squirrels' (see Fig. 3.7) weight and size are affected by shorter winters. Even though this looks

Fig. 3.7 Arctic ground squirrels (*Urocitellus paryii*)

Fig. 3.8 A western hedgehog (*Erinaceus europaeus*)

unfavorable, it can be even worse when considering that this impression is caused by comparing the females that survive longer winters with the females that survive shorter winters. The females that survive longer winters may seem stronger, simply because the weaker females did not survive the longer winter. This would mean that weaker female squirrels can also reproduce after shorter winters, which in turn affects the strength of the species.

The opposite is true for Western hedgehogs (see Fig. 3.8). They have better body conditions after shorter winters, as they lose less weight over the hibernation period. Also, several bat species such as pallid bats (see Fig. 3.9) benefit from warmer temperatures. This is because they need less energy during hibernation and more insects are available as prey when they wake up.

Fig. 3.9 A pallid bat (*Antrozous pallidus*)

Fig. 3.10 A common hamster (*Cricetus cricetus*)

3.3.3   Reproduction

The third way changing climatic circumstances can impact hibernating animals is by affecting their reproduction, such as the number of babies. For example, common hamsters (see Fig. 3.10) have twice as many babies in warmer, shorter winters. Also, more male rodents can have offspring with warmer temperatures. This is

Fig. 3.11 Uinta ground squirrel (*Urocitellus Armatus*)

because they need to keep a high body temperature for several weeks to become sexually mature. The warmer it is, the easier it is to keep their body temperature high.

3.3.4 Survival

The fourth way changing climatic circumstances can impact hibernating animals is by affecting their survival. Higher temperatures can make survival more but also less likely. Or make no difference at all, which means that survival very much depends on several factors. These factors include species, seasonality, and age.

For example, warmer temperatures make the survival of young Uinta ground squirrels (see Fig. 3.11) more likely than the survival of 1-year-olds and adults. This is probably because the young squirrels need more energy to stay warm during hibernation.

Also, with more frost in fall, adult alpine marmots (see Fig. 3.12) are less likely to survive hibernation. This is probably because colder temperatures in fall require them to use more energy, meaning less is available for the hibernation phase.

3.4 Conclusion

So, climate change not only affects the environment and us, it also affects animals in different ways. These impacts are not always obvious. For example, changing environmental circumstances influence the number of female turtles because

Fig. 3.12 Alpine marmot (*Marmota marmota*)

warmer temperatures increase the nest temperature and a higher nest temperature causes more females to be born.

Also, changing environmental circumstances in summer impacts honey bees' survival in winter because too much cold or heat and too much or little rain impact how well plants and flowers can grow. This in turn impacts whether honey bees can gather enough high-quality pollen and nectar for the winter. And summers that are too long help varroa mites to thrive, which affects the health of honey bee colonies.

Moreover, global warming can impact hibernating animals by affecting their phenology, condition, reproduction, and survival. But not all animals are affected equally, as the impact for example depends on species, age, and condition.

3.5   How We Can Take Action

As climate change is caused by human activities, it is helpful to limit greenhouse gas emissions, such as CO_2 emissions as much as possible. Here are practical ideas of what you and I can do to mitigate climate change by limiting CO_2 emissions:

- Reducing the amount of meat and dairy in meals
- Choosing environmentally friendly transportation methods, such as taking a train instead of flying or cycling instead of driving a car
- Choosing renewable energy sources with energy provider
- Using environmentally friendly building materials when building a house or office

Here are practical ideas of what you and I can do to extract CO_2 from the atmosphere:

- Planting trees
- Donating to an organization that engages in carbon offset projects
- Investing in technologies that can remove CO_2 from the atmosphere
- Planting cover crops on empty land instead of keeping the land bare so that more CO_2 can be stored in the soil
- Using kerosene that is made from water, CO_2, and sunlight instead of kerosene made from fossil fuel (further reading: Chap. 6: "Climate Solutions: Reducing CO_2 Emissions in the Aviation Sector")

Credit

This Chapter Is Based On:

Sea Turtles:
Laloë, J. O., Tedeschi, J. N., Booth, D. T., Bell, I., Dunstan, A., Reina, R. D., & Hays, G. C. (2021). Extreme rainfall events and cooling of sea turtle clutches: Implications in the face of climate warming. *Ecology and Evolution, 11*(1), 560–565.

Honey Bees:
Calovi, M., Grozinger, C. M., Miller, D. A., & Goslee, S. C. (2021). Summer weather conditions influence winter survival of honey bees (Apis mellifera) in the northeastern United States. *Scientific Reports, 11*(1), 1–12.
Wilson Rankin, E. E., Barney, S. K., & Lozano, G. E. (2020). Reduced water negatively impacts social bee survival and productivity via shifts in floral nutrition. *Journal of Insect Science, 20*(5), 15.

Hibernating Animals:
Wells, C. P., Barbier, R., Nelson, S., Kanaziz, R., & Aubry, L. M. (2022). Life history consequences of climate change in hibernating mammals: A review. *Ecography, 2022*(6), e06056.

Figure Credits

Fig. 3.1   Kalaeva on Shutterstock
Fig. 3.2   Mirko Graul on Shutterstock
Fig. 3.3   Owsigor on Shutterstock
Fig. 3.4   Fremme on Shutterstock
Fig. 3.5   Debbie Steinhausser on Shutterstock
Fig. 3.6   Diego Cottino on Shutterstock
Fig. 3.7   Jukka Jantunen on Shutterstock
Fig. 3.8   Mr. SUTTIPON YAKHAM on Shutterstock
Fig. 3.9   Danita Delimont on Shutterstock
Fig. 3.10 An13nA on Shutterstock
Fig. 3.11 Liz Weber on Shutterstock
Fig. 3.12 Astrid Gast on Shutterstock

Chapter 4
How Climate Change Impacts Pregnant Women and Birth Defects

Abstract The effects of climate change are not always visible, but they still have significant consequences, including on pregnant women and birth outcomes. These consequences can be noticeable during pregnancy but also impact children throughout their lives. Issues during pregnancy involve high blood pressure, gestational diabetes, preterm delivery, spontaneous abortion, and lower birth weight. Other issues that may arise during pregnancy and can lead to lifelong problems include facial clefts, hypospadias, autism, and heart defects. Even our society can be impacted when sex ratios change.

Keywords Science · Science communication · Climate change · Climate change consequences · Global warming · Health · Pregnancy · Newborns · High blood pressure · Gestational diabetes · Preterm delivery · Spontaneous abortion · Birth weight · Hypospadias · Orofacial clefts · Autism · Heart defects · Sex ratio · Birth outcome · Perinatal period · Pregnancy outcome

While some of the consequences of climate change are obvious and well-known, such as changing mountains and mountain ecosystems, other consequences of climate change are not obvious and not well-known. For example, most of us do not know that rising temperatures and changing humidity are also slowly changing the physiology of our physical bodies. Physiology concerns the normal functions of living organisms and the parts they consist of.

For example, the human body contains several mechanisms that make sure we have the right temperature independent of external temperatures. When outside temperatures become extremely hot or cold, the body has to work harder to keep this right temperature. This is stressful. Initially, this stress doesn't cause any problems, but in the long term, it can affect many body systems and cause all kinds of problems, including heart and blood vessel problems, and lung and breathing problems. And eventually, it can cause us to die.

Credit: This chapter is based on the scientific article "Impacts of climate changes on pregnancy and birth outcomes: A review" by Bahare Dehdashti and colleagues. (Full citation is available at the end of the chapter).

Fig. 4.1 Very hot temperatures can cause all kind of problems for both mum and baby

The stress caused by extreme temperatures is especially critical in more vulnerable groups, such as elderly people and people with disabilities. And also pregnant women and their fetuses are at risk (see Fig. 4.1). This is how higher temperatures caused by climate change affect pregnancy and birth outcomes:

4.1   High Blood Pressure

The first way higher temperatures caused by climate change affect pregnancy and birth outcomes is by causing high blood pressure. This is because high temperatures affect the placenta's blood vessels at the beginning of the pregnancy and increase stress levels at the end of the pregnancy. The consequences can be:

- hypertension: regular high blood pressure
- preeclampsia: high blood pressure and organ damage
- eclampsia: a seizure or coma caused by high blood pressure

These problems can in turn cause further issues, such as the placenta separating itself from the uterus, failing organs, and the fetus' death.

4.2   Gestational Diabetes

The second way higher temperatures caused by climate change affect pregnancy and birth outcomes is by increasing the likelihood of gestational diabetes. Gestational diabetes is carbohydrate intolerance during pregnancy. Hot summers make this type of diabetes more likely to occur because high temperatures affect the activity of the pancreas, which is the organ that produces insulin to regulate blood sugar levels. If not treated, meaning the mom's blood sugar levels are not regulated, it increases the number of ill babies or babies who die during pregnancy. This is because both the baby and the mother will have higher blood sugar levels, causing the baby to produce excessive amounts of insulin. This can in turn cause insulin resistance which affects glucose tolerance.

4.3   Preterm Delivery

The third way higher temperatures caused by climate change affect pregnancy and birth outcomes is by increasing the likelihood of preterm delivery (see Fig. 4.2). Preterm delivery is delivery before the 37th week of pregnancy. It affects the baby's health, causes for example behavioral, social, and neurodevelopmental disorders, and can cause communication and learning problems.

Apart from higher environmental temperatures, also other extreme environmental circumstances caused by climate change can cause preterm deliveries. For example, storms and floods can increase the risk of preterm delivery.

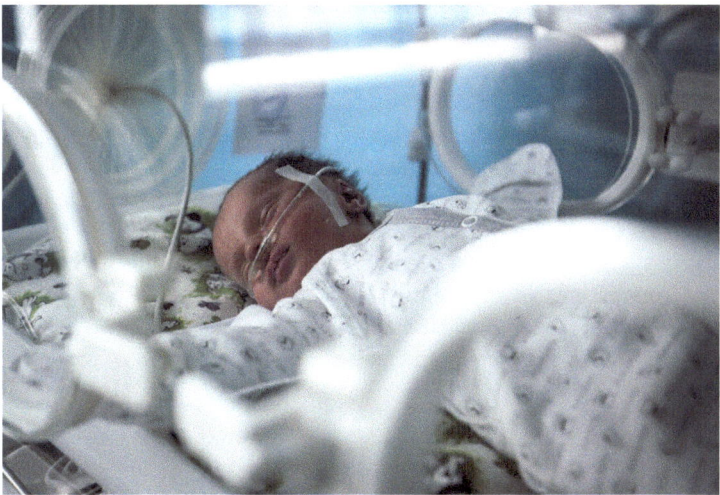

Fig. 4.2 Environmental changes cause an increase in premature delivery rates

4.4   Spontaneous Abortion

The fourth way higher temperatures caused by climate change affect pregnancy and birth outcomes is by increasing the likelihood of spontaneous abortion. Spontaneous abortion is the natural loss of the baby before the twentieth week of pregnancy. This is because high global temperatures can affect the hormone levels in pregnant women, their blood pressure, and the amounts of blood that is circulating from the mother to the fetus. Also, other environmental changes, such as natural disasters, can cause a lot of stress. This stress can cause a miscarriage or the baby's death during pregnancy or delivery.

4.5   Birth Weight

The fifth way higher temperatures caused by climate change affect pregnancy and birth outcomes is by impacting birth weight. This is because high environmental temperatures can cause low birth weight. Low birth weight is less than 2.5 kg (5.5 lbs) for a baby born between 37 and 44 weeks of pregnancy.

Birth weight is an indicator of fetal growth. Limited fetal growth can be caused directly by high or changing temperatures, and indirectly by for example infections in the mother's body, her psychological problems, and limited food availability. This can cause problems with the placenta, fetus, and vascular system and in turn health problems.

4.6   Hypospadias

The sixth way higher temperatures caused by climate change affect pregnancy and birth outcomes is by increasing the likelihood of hypospadias. Hypospadias is a malformation of male children's sex organs so that the urethral opening is not at the tip of the penis (see Fig. 4.3). Normally, about 1 in every 200–300 boys has this condition, but this number is increasing due to environmental influences during pregnancy. Apart from preventing normal urine flow, it can also cause infertility. Also, even when this is repaired, it can continue to cause issues such as a shorter penis, curving of the penis, leaks, and pain during sex.

4.7   Orofacial Clefts

The seventh way higher temperatures caused by climate change affect pregnancy and birth outcomes is by increasing the likelihood of orofacial clefts. Orofacial clefts occur when a baby's lip or mouth does not form properly during pregnancy

Types of hypospadias

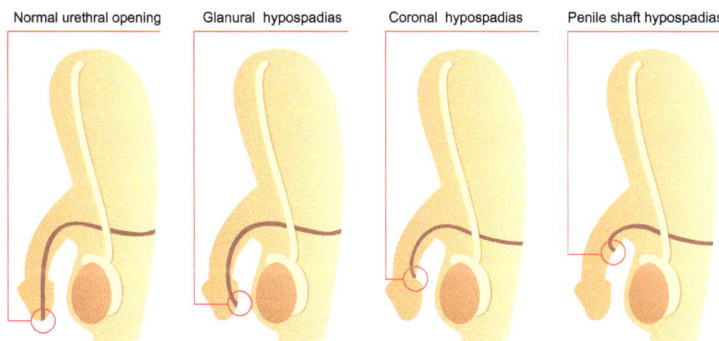

Fig. 4.3 Hypospadias is a condition in which the urethral opening is not at the tip of the penis

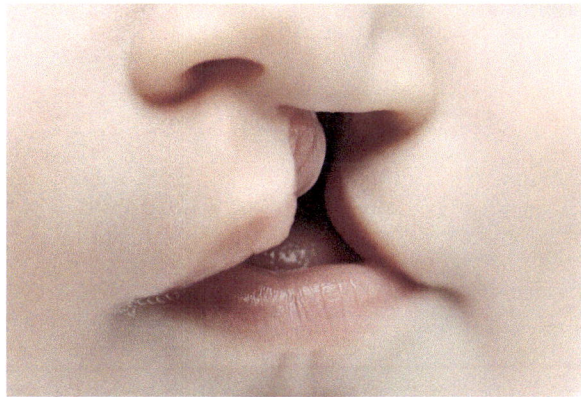

Fig. 4.4 A baby with an untreated facial cleft

(see Fig. 4.4). Babies with oral clefts often suffer because of difficulties in feeding, hearing, and speaking, and also often have dental and psychological problems. Even though the causes of oral clefts are not completely clear, it is clear that both genetic and environmental factors such as higher temperatures play a role in their development.

4.8   Autism

The eighth way higher temperatures caused by climate change affect pregnancy and birth outcomes is by increasing the likelihood of developing autism. Autism is a developmental condition caused by differences in the brain. It is characterized by for example an unusual interest in objects, difficulties with changes in daily routines, great ability in one area and great difficulty in another area, and unusually strong reactions to input to one of their five senses. About 1 in 88 children has autism, caused both by genetic and environmental factors. Environmental factors such as high temperatures can also impact which genes are expressed and to what extent, without changing the DNA sequence.

4.9   Heart Defects

The ninth way higher temperatures caused by climate change affect pregnancy and birth outcomes is by increasing the likelihood of heart defects at birth. These heart defects can both be structural and functional and are the main cause of infant mortality. Some causes of heart defects are hereditary but also triggered by environmental factors such as high temperatures, especially during the first 3 months of pregnancy. It can cause ventricular septal defects, which is a hole in the wall between the two lower compartments of the heart (see Fig. 4.5), and atrial septal defects, which is a hole between the top two compartments of the heart.

4.10   Sex Ratios

The tenth way higher temperatures caused by climate change affect pregnancy and birth outcomes is by affecting sex ratios. The normal sex ratio is around 1050 boys per 1000 girls (see Fig. 4.6). With every 1 °C (1.8 °F) increase in annual temperature, this ratio changes to 1051 per 1000 girls. This means that higher temperatures

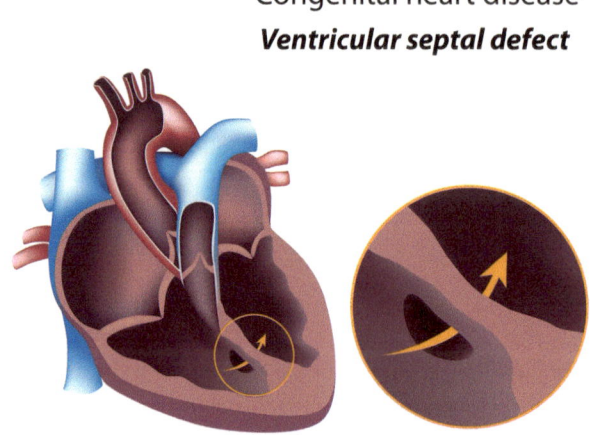

Congenital heart disease
Ventricular septal defect

Fig. 4.5 A ventricular septal defect is a hole in the wall between the heart's lower chambers

Sex ratio at birth, 2017
Sex ratio at birth, measured as the number of male births per 100 female births. Birth ratios are slightly male-biased,
with an expected biological ratio of 105 male per 100 female births.

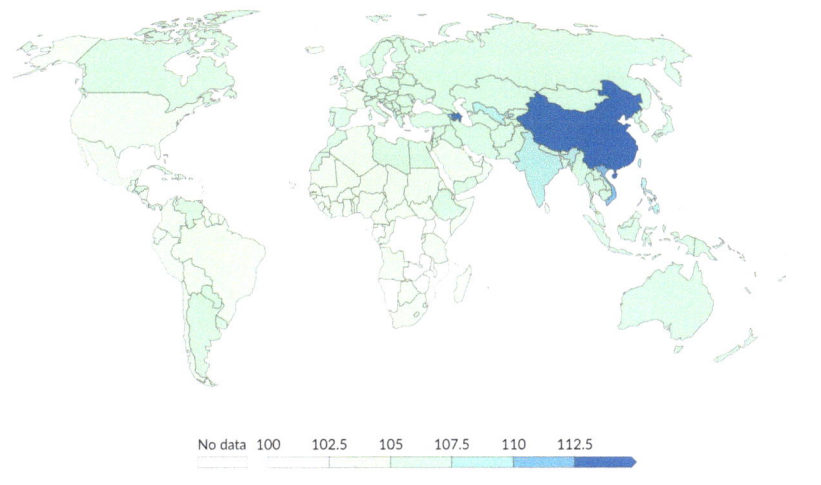

Data source: Chao et al. (2019)   OurWorldInData.org/gender-ratio | CC BY

Fig. 4.6   Sex ratios at birth in different parts of the world

cause more boys to be born—as opposed to sea turtles, as higher temperatures cause more female turtles to be born (see Sect. 3.1). This increase of 1 may sound like a tiny difference, but with a global population of 8 billion people, this means about 3.8 million more men with a 2 °C (3.6 °F) increase in global temperature.

4.11   Conclusion

So, even when the effects of climate change are not always visible, the possible consequences on pregnant women and unborn babies are very serious. This is because these consequences are not always limited to the nine-month pregnancy period.

Issues that mainly impact the mother and baby during pregnancy are high blood pressure, gestational diabetes, preterm delivery, spontaneous abortion, and lower birth weight. Other issues can usually be repaired after birth, such as facial clefts and hypospadias, although they will always be visible and can continue to cause problems. But some issues will impact a child's whole life, such as autism and heart defects. Or even society when sex ratios change.

4.12   How We Can Take Action

As climate change and resulting higher temperatures can have such a huge impact on the health of women and our children, we need to do as much as we can to limit climate change. Here are practical ideas of what you and I can do to prevent and limit climate change when having a baby:

- Eating healthy, organic food
- Buying second-hand baby clothes and other products
- Using reusable instead of single-use diapers
- Planting a tree when a baby is born
- Breastfeeding a baby if possible
- Buying sustainable baby products
- Visiting nature to allow the child to connect with nature from an early age

Credit

This Chapter Is Based On:

Dehdashti, B., Bagheri, N., Amin, M. M., & Hajizadeh, Y. (2020). Impacts of climate changes on pregnancy and birth outcomes: A review. *International Journal of Environmental Health Engineering*, 9(1), 24.

Figure Credits

Fig. 4.1   Onjira Leibe on Shutterstock
Fig. 4.2   home for heroes on Shutterstock
Fig. 4.3   Irinaloveliness on Shutterstock
Fig. 4.4   malost on Shutterstock
Fig. 4.5   Alila Medical Media on Shutterstock
Fig. 4.6   "Gender Ratio" by Chao et al. Is licensed under CC BY 4.0 DEED
           Source: https://ourworldindata.org/gender-ratio#sex-ratio-at-birth
           Author: Chao, F., Gerland, P., Cook, A.R., Alkema, L. (2019). Systematic assessment of
                   the sex ratio at birth for all countries and estimation of national imbalances and
                   regional reference levels. PNAS.
           License: https://creativecommons.org/licenses/by/4.0/

Chapter 5
Climate Solutions: Reducing CO_2 Emissions in the Building Sector

Abstract As climate change has significant consequences on our health and the environment, it is crucial to reduce the negative impact of our activities. For example, reducing the negative impact in the building sector is helpful because this sector uses 10–15% of total industrial energy and causes 5–8% of the human CO_2 emissions. Emissions in this sector can be limited by reducing direct emissions, for example by replacing conventional by self-healing bioconcrete. We can also reduce indirect emissions by, for example, installing heat-absorbing walls that use solar energy for indoor heating. Also, we can reduce other indirect emissions, for instance by using natural materials such as bamboo instead of synthetic materials.

Keywords Science · Science communication · Climate change · Climate change solutions · CO_2 emissions · Greenhouse gas · Building sector · Sustainable materials · Bioconcrete · Durability · Bamboo · Trombe wall · Energy efficiency · Fungi · Concrete repair · Cracks · Crack healing · Passive energy systems · Solar thermal energy · Passive heating · Mechanical properties · Mechanical engineering · Innovation · Energy

Climate change has many serious consequences, both directly on for example our health, and indirectly for example by affecting our environment. This is why it is critical that we reduce the negative impact of our activities.

One sector in which our activities strongly contribute to climate change is the building sector. This is because concrete is a commonly used building material: over 30 million tons of concrete are being produced and used every year. This is about the weight of six Pyramids of Giza. Unfortunately, the production of cement, which is a main component of concrete, is environmentally unfriendly: it uses 10–15% of total industrial energy and causes 5–8% of the CO_2 emissions caused by humans.

Credit: This chapter is based on five scientific articles by Chereddy Sonali Sri Durga, Carolina Martuscellia, Gehad Metwally, Piotr Borowski, and Aleksejs Prozuments, and their colleagues. (Full citations are available at the end of the chapter)

Because of this significant contribution to climate change, it is helpful to make the building industry more environmentally friendly by reducing CO_2 emissions. This can be done in different ways:

- by reducing direct emissions, for example during the production process
- by reducing indirect emissions caused by the generation of energy, for example when electricity is created by burning coal
- by reducing other indirect emissions that are caused by other parts of the supply chain, so before or after a company's activities, for example reducing emissions caused by waste treatment

Here are examples of how this can be put into practice:

5.1   Reducing Direct Emissions: Using Self-Healing Concrete

The first way to reduce CO_2 emissions in the building sector is by reducing direct emissions. This can for example be done by using self-healing concrete. Self-healing concrete is concrete that can restore cracks without being replaced or repaired. This is helpful because conventional concrete needs to be replaced and repaired more often as it easily cracks. Cracks can appear when the temperature changes and the concrete expands or contracts. Gases such as air and liquids such as rain can flow in, causing the concrete structure to decay. Not only does the concrete not last as long but also the metal inside the concrete can rust. This means that (micro-)cracks can cause structural failure and make the building less durable and safe.

Structural failure is harder to prevent with conventional concrete, as it cannot heal itself. But bioconcrete can! Bioconcrete is concrete that contains bacteria in the form of spores. When these spores get into contact with water because water flows in a crack, they become active and produce crystals made of calcium carbonate (see Fig. 5.1). Calcium carbonate is a well-known substance, as eggshells mostly consist of it and it is used as a meal supplement to treat calcium deficits. The number of crystals produced by these bacteria grows until the whole crack is closed.

Closing cracks with calcium carbonate is environmentally friendly, not only because it is natural but it also helps reduce CO_2 emissions. CO_2 emissions are reduced, because:

- The life span of concrete structures increases, so less concrete needs to be produced. As mentioned before, producing cement causes a lot of CO_2 emissions.
- The concrete does not need inspection and restoration, which also requires material and energy.
- CO_2 is consumed in the calcium carbonate: carbon dioxide and calcium hydroxide (slaked lime) form calcium carbonate and water, as can be seen in the chemical reaction $CO_2 + Ca(OH)_2 \rightarrow CaCO_3 + H_2O$.

Fig. 5.1 Concrete that has healed itself

5.2   Reducing Indirect Emissions: Using Trombe Walls

The second way to reduce CO_2 emissions in the building sector is by reducing indirect emissions. This can for example be done by using heat-absorbing walls, such as walls that use Trombe wall technology (see Fig. 5.2). In the video in Fig. 5.3, you can see an example of a Trombe wall in a house, both from the outside and inside.

This is why heat-absorbing walls using the Trombe wall technology are a great way to reduce CO_2 emissions:

5.2.1   Simple and Maintenance-Free Solution

The first reason why heat-absorbing walls are a great way to reduce CO_2 emissions is that it is a maintenance-free solution. This is because it is a simple technology once it is set up: these walls only consist of a solid wall, a glass or other clear outer plate, and an air gap in between. The wall is usually painted dark so that more heat from solar energy can be collected and stored. The goal of the glass plate is to make the air in the air channel more likely to warm up. The stored energy in the wall is then transferred to the other side of the wall, which is typically a room, so that this room is heated.

While the overall setup is simple, the details may be different and can be more complex or easier depending on the specific geographic and climatic circumstances

Fig. 5.2 Example of a Trombe wall

Fig. 5.3

and the purpose of the building. For example, the wall system needs to take both extremely high and low temperatures into account so that the system doesn't get damaged. And walls used for factories can be simpler than for houses, as indoor comfort is unimportant in factories.

As this technology is so simple and mainly consists of glass and masonry, the CO_2 emissions produced during production are limited. Also, as the required materials are common, it is more likely that they can be produced locally. This reduces CO_2 emissions caused by transport. And as no maintenance is required of a wall, no CO_2 is emitted as soon as the wall is operational.

Fig. 5.4

In the video in Fig. 5.4, you can see a simulation of how a Trombe wall heats the air in a room.

5.2.2   Energy Reduction for Heating and Cooling

The second reason why heat-absorbing walls are a great way to reduce CO_2 emissions is that they reduce the amount of energy needed for heating and even cooling.

The amount of energy needed for heating is reduced because the wall allows air to warm up by solar energy during the day. This heat is stored in the wall and slowly radiated into the colder room. Also, warm air is automatically sucked into the colder room because of differences in air pressure and can be sped up using a fan. The image in Fig. 5.5 shows how it works: cold air has a higher air pressure so wants to move out of the room, through a valve, into the Trombe wall. Here, it is heated by the sun and flows through another valve back into the room. As the sun is not a stable energy source, this system cannot be used as the main heating, but it helps us use no or less heating from additional sources, such as gas, wood, or electricity. This reduces the CO_2 emissions from burning fossil fuels and wood.

The amount of energy needed for cooling is also reduced because the same setup can be used for cooling the house. This is very helpful because heat from solar radiation is especially available in warm seasons. But in warm seasons, cooling instead of heating is more important. The image in Fig. 5.6 shows how it works: the valves in the walls are closed, so that the cold air stays in the room and the warm air stays outside. The stacking or chimney effect caused by differences in air pressure causes the air to flow past the wall, cooling down the wall. As a consequence, less energy is needed for air-conditioning.

5.2.3   Renewable Energy Source

The third reason why heat-absorbing walls are a great way to reduce CO_2 emissions is that heating is based on a renewable energy source. A renewable energy source is an energy source that is replenished relatively quickly, within a human lifetime. Solar, wind, and wave energy are great examples. This is as opposed to

Fig. 5.5 During colder seasons, cold air (*blue arrow*) flows into the wall and flows back as heated air (*red arrow*)

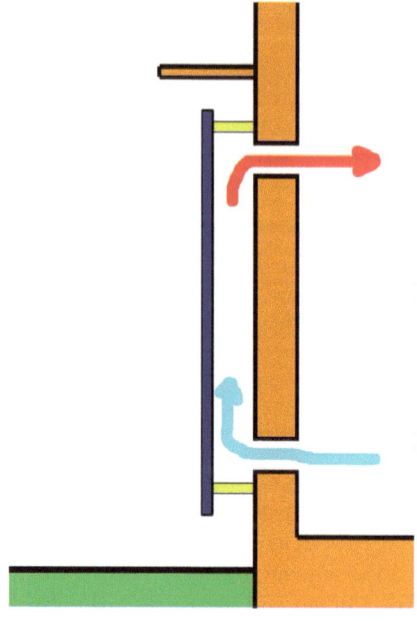

Fig. 5.6 Warm air (*red arrows*) can be kept outside in summer

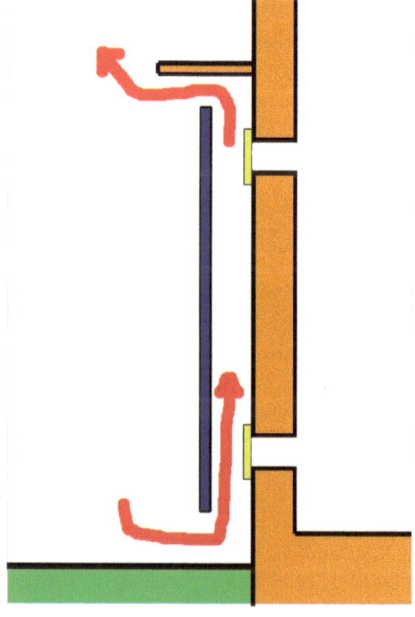

non-renewable energy sources that take millions of years to be replenished. An example is fossil fuels such as coal, gas, and crude oil.

As solar energy is a renewable energy source, this means that no CO_2 emissions are caused by operating this technology. But this also means that these walls are only effective in sunny areas: in cloudy or overcast areas, not enough heat is provided by solar energy. Still, it can be effective to for example preheat air. Also, the technology can be extended by adding a component that stores heat for a longer time, so that heat is available even when the amount of sunshine is inconsistent. Storage can for example be achieved by using materials that change from solid to liquid when heat from the sun is available, and turn back into a solid when the heat is radiated into the room.

5.3 Reducing Other Indirect Emissions: Using Natural Materials

The third way to reduce CO_2 emissions in the building sector is by reducing other indirect emissions. This can for example be done by replacing synthetic with natural materials. Synthetic materials have in the last few decades been developed to create materials with beneficial characteristics. But while they have many advantages, they also have important disadvantages. For example, many of these materials, such as fossil-fuel-based plastics, cause a lot of CO_2 emissions throughout their lifecycle. Also, they are not biodegradable, which means they can cause long-term pollution in the environment (further reading: Chap. 7 of A Guide to a Healthier Planet Volume 1: "How Plastic Pollution Impacts Our Environment") and require a lot of energy for recycling.

As synthetic materials are usually an environmentally unfriendly choice, the focus in the current efforts to use earthly resources more sustainably in the building sector is moving back to natural materials. One great sustainable material is bamboo (see Fig. 5.7) because it benefits the environment throughout its lifespan.

Bamboo helps mitigate climate change throughout its lifespan especially during growth because it removes—just like other plants—CO_2 from the air as it grows. But—in contrast to other plants—bamboo has many amazing characteristics that make it even better at doing so.

One amazing characteristic is that bamboo culms have a shared root system that spreads quickly. A culm is the bamboo's stem. It is called a stem as opposed to a trunk, as bamboo is a type of grass, not a tree. As bamboo spreads easily, a bamboo forest develops quicker than a tree forest. As a consequence, it can extract more CO_2 from the atmosphere sooner.

Another amazing characteristic is that bamboo grows fast. Depending on the species, it can grow almost 1 m (3 ft) per day! It increases its mass by 10–30% percent a year, which is much more than trees (2–5%). Also, bamboo grows to maturity in only 3–5 years. This quick growth cycle makes it possible to capture three times more CO_2 than trees.

Fig. 5.7 Bamboo (*Bambusa*) is a great, sustainable material

Fig. 5.8

A third amazing characteristic is that bamboo can thrive in a wide range of environments, including tropical, subtropical, and temperate regions. It can even grow in the freezing cold (-20 to -15 °C / -4 to 5 °F) and lands that are otherwise unproductive such as ravines.

This combination of spreading and growing quickly, and being able to grow in a wide range of environments makes bamboo a suitable strategy for climate change mitigation. It is estimated that planting an additional 100,000 km^2 (about 39,000 mi^2; which is about the size of South Korea) on degraded lands, bamboo, and bamboo products can save over 7 gigatons of CO_2 within 30 years. To compare, this is equivalent to the CO_2 300 million electric cars would save compared to fossil-fuel-based cars in the same time.

While this shift seems to be creative, the use of bamboo is not new: it has been used in Asia already for a very long time as they recognized it as a very suitable material for construction. And while unprotected bamboo is eaten by insects and degrades relatively quickly, causing structures to fail, new technologies can protect it, allowing us to use bamboo for many more purposes. The video in Fig. 5.8 shows examples of how bamboo can be used in the building sector and the gorgeous result.

5.4 Conclusion

So, as the building sector causes a lot of CO_2 emissions, it is important to reduce these emissions. This can be done by reducing direct and indirect emissions.

Reducing direct emissions can for example be achieved by using self-healing concrete. This is helpful because self-healing concrete lasts longer than conventional concrete and as a consequence has to be replaced less often.

Reducing indirect emissions can for example be achieved by building heat-capturing Trombe walls. These walls use energy from sunlight to heat indoor spaces, reducing the need to burn fossil fuels to generate heat. They can even be used for cooling! Or it can for example be achieved by replacing synthetic by natural materials, such as bamboo.

And more good news: the positive effects are often even larger than we realize because CO_2 emissions are usually reduced in several ways at once. For example, when using trombe walls, other indirect emissions are reduced as well because locally produced materials can be used, which reduces emissions caused by transportation.

5.5 How We Can Take Action

As reducing CO_2 emissions in the building sector can make a significant difference to limiting climate change, here are practical ideas of what you and I can do to slow down climate change:

- Using self-healing instead of conventional concrete for buildings
- Using a Trombe wall for space heating and cooling
- Insulating the walls and roof of your house
- Reducing the temperature to which a room is heated up
- Using bamboo in construction
- Using bamboo for indoor structures, furniture, and other objects we use in daily life
- Using bamboo ash to improve the properties of cement

Credit

This Chapter Is Based On:

Self-Healing Concrete:
Durga, C. S. S., Ruben, N., Chand, M. S. R., & Venkatesh, C. (2020). Performance studies on rate of self healing in bio concrete. *Materials Today: Proceedings, 27*, 158–162.

Martuscelli, C., Soares, C., Camões, A., & Lima, N. (2020). Potential of fungi for concrete repair. *Procedia Manufacturing*, *46*, 180–185.
Metwally, G. A., Mahdy, M., & El-Raheem, A. H. (2020). Performance of bio concrete by using Bacillus Pasteurii bacteria. *Civil Engineering Journal*, *6*(8), 1443–1456.

Bamboo:
Borowski, P. F., Patuk, I., & Bandala, E. R. (1955). Innovative industrial use of bamboo as key "green" material. *Sustainability*, *2022,* 14.

Trombe Walls:
Prozuments, A., Borodinecs, A., Bebre, G., & Bajare, D. (2023). A review on Trombe wall technology feasibility and applications. *Sustainability*, *15*(5), 3914.

Figure Credits

Fig. 5.1   studiojaskrawo on Shutterstock
Fig. 5.2   "Passive Solar Trombe-wall" by PSHiker is licensed under CC BY-NC-SA 2.0
           Source: https://flic.kr/p/pDBqA7
           Author: https://www.flickr.com/photos/pshiker/
           License: https://creativecommons.org/licenses/by-nc-sa/2.0/
Fig. 5.5   "Trombe" by juanantonaya us published in the public domain
           Source: https://commons.wikimedia.org/wiki/File:Trombe1.png
           Author: https://commons.wikimedia.org/wiki/User:Juanantonaya
           License: https://en.wikipedia.org/wiki/Public_domain
Fig. 5.6   "Trombe" by juanantonaya is published in the public domain
           Source: https://commons.wikimedia.org/wiki/File:Trombe4.png
           Author: https://commons.wikimedia.org/wiki/User:Juanantonaya
           License: https://en.wikipedia.org/wiki/Public_domain
Fig. 5.7   Sofiaworld on Shutterstock

Chapter 6
Climate Solutions: Reducing CO_2 Emissions in the Aviation Sector

Abstract The aviation industry contributes significantly to global CO_2 emissions, accounting for about 2% of the global CO_2 emissions. To make the sector more environmentally friendly, CO_2 emissions can be reduced by using kerosene made from CO_2, water, and solar energy instead of using fossil fuel-based kerosene. Or other more environmentally friendly fuels can be used, such as hydrogen or biofuel. Even electric planes exist. Also, the aviation sector can be made more environmentally friendly by improving airplane design and technologies, making operations on the ground and in the air more efficient, and by carbon offsetting CO_2 emissions.

Keywords Science · Science communication · Climate change · Climate change solutions · CO_2 emissions · Greenhouse gas emissions · Aviation sector · Kerosene · Aviation fuel · Sustainable air transportation · Concentrated solar energy · Thermochemical solar fuels · Solar reactor · Redox cycle · Water splitting · CO_2 splitting · Ceria · Sustainable aviation fuels · Energy consumption · Mitigation · Sustainable · Aviation fuel

Another industry that causes a lot of CO_2 emissions is the aviation industry. This is because air transportation of both people and cargo benefits the world in so many ways that it has become an important aspect of modern daily life. For example, it allows couples who originate from different parts of the world to see their families and it contributes to economic growth.

While the number of people traveling and cargo shipped by plane increases, so does the negative impact of air transportation on the environment: with more planes flying to more places, the amount of fuel consumption increases. As the aviation industry relies on fossil fuels, greenhouse gasses are emitted during combustion, mostly CO_2: air transportation worldwide contributed 915 million tons of CO_2 in 2019, accounting for 2% of the global CO_2 emissions. This is about the same amount of CO_2 that all humans on Earth breathe out in 4 months.

Credit: This chapter is based on two scientific articles by Md Arif Hasan and Stefan Zoller and their colleagues. (Full citations are available at the end of the chapter)

As the negative impact of CO$_2$ on our climate will increase with more air transportation, it is important to make the aviation sector more environmentally friendly. This is how:

6.1   Environmentally Friendly Kerosene

A first way to make the aviation sector more environmentally friendly is by using environmentally friendly kerosene. This is helpful because kerosene is currently the main fuel, causing most of the CO$_2$ emissions.

Typically, kerosene is made from fossil fuels by heating crude oil to 160–250 °C (320–482 °F). The same process is used to create petrol and diesel, but just at a different temperature (see Fig. 6.1).

But a relatively recent solution allows us to create environmentally friendly kerosene, made from CO$_2$ and water molecules with the help of sunlight, a renewable energy source. This kerosene can replace kerosene created from fossil fuels on an industrial scale while using existing global jet fuel infrastructures and engines! This is how kerosene can be created using CO$_2$, water, and solar energy:

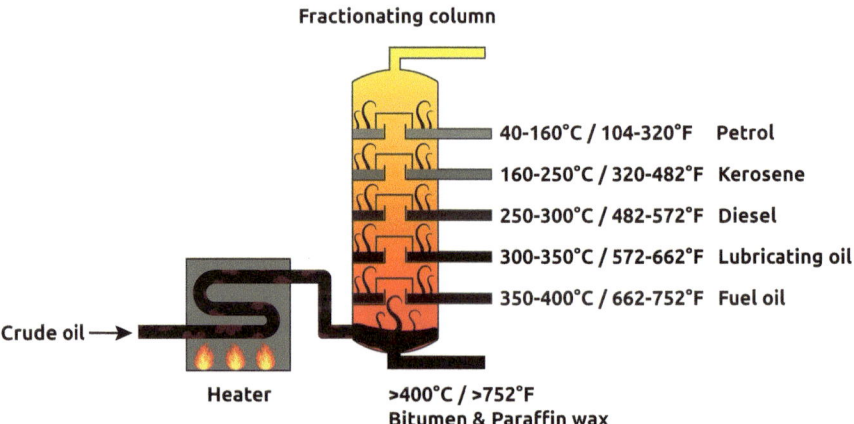

Fig. 6.1 Different types of fuel are created by heating crude oil to different temperatures

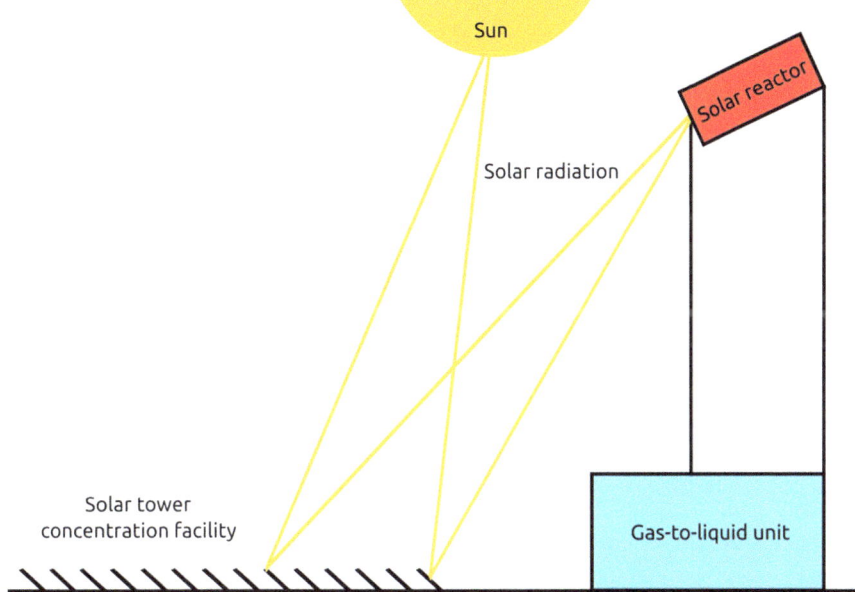

Fig. 6.2 The facility used to produce kerosene consists of a solar tower concentration facility, a solar reactor, and a gas-to-liquid unit

6.1.1   Setup

To be able to create kerosene from CO_2, water, and solar energy, a facility is used that consists of three main parts (see Fig. 6.2).

The first part is the solar tower concentration facility. This facility is located on the ground and consists of 169 reflectors in 14 rows. These reflectors are not flat like mirrors, but slightly spherical like a dish for satellite reception, so that the sunlight is concentrated in one point. This point is located at the top of the tower in the solar reactor. By bringing so many sunbeams together, the energy gathered at the top of the tower is about the same as 2500 suns and at peak times above 4000 suns!

The second part is the solar reactor. This reactor is located 15 m (49.2 ft) above the ground and tilted by $40°$ so that the sunbeams from the ground can easily enter. In this reactor, gases are produced with the help of the energy from the sunlight. These gases leave the reactor at the back and flow down toward the gas-to-liquid unit.

The third part is the gas-to-liquid unit. This unit is located in a container on the ground next to the solar tower. Here, the gases produced in the solar reactor are converted into kerosene.

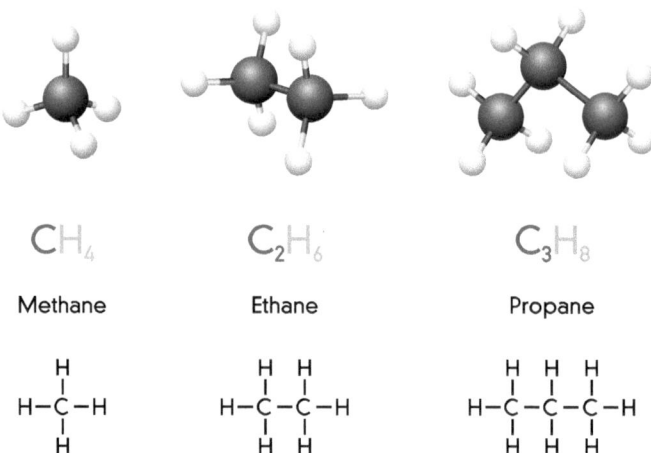

$$CH_4$$

Methane

$$C_2H_6$$

Ethane

$$C_3H_8$$

Propane

```
     H                H  H            H  H  H
     |                |  |            |  |  |
 H - C - H        H - C - C - H   H - C - C - C - H
     |                |  |            |  |  |
     H                H  H            H  H  H
```

Fig. 6.3 Example hydrocarbons, which can increase in size, depending on the number of carbon atoms

6.1.2   How It Works

The facility consisting of the solar tower concentration facility, solar reactor, and gas-to-liquid unit creates kerosene in several steps.

The first step in the process of creating kerosene is transforming sunlight energy into syngas. Syngas is a mixture of hydrogen (H_2) and carbon monoxide (CO) gases. This syngas is produced from water (H_2O) and carbon dioxide (CO_2) molecules in the air by letting these molecules chemically react with ceria under high temperatures (900 °C/1652 °F). Ceria is a type of ceramics.

The second step in the process of creating kerosene is letting the produced syngas flow into a buffer tank in the gas-to-liquid unit. Here it can be stored under pressure until it is used for the third step.

The third step in the process of creating kerosene is converting the syngas into kerosene. Here, chemical reactions take place that rearrange the atoms of the H_2 and CO gases into hydrocarbons. A hydrocarbon is a molecule consisting of only hydrogen (H) and carbon (C) atoms (see Fig. 6.3). For example, a small hydrocarbon molecule is ethane, consisting of 2 carbon atoms and 6 hydrogen atoms. Hydrocarbons in kerosene are larger and usually contain between 11 and 15 carbon atoms per molecule. These chemical reactions take place at a lower temperature (210 °C/410 °F), but higher pressure (30 bar; to compare that is about 15 times the pressure of a car tire).

The fourth and last step in the process of creating kerosene is reducing the pressure and temperature of the output of these chemical reactions. The result consists of liquid, wax, and gas. 16% of the liquid and 7% of the wax is kerosene. Any gases that are left are recycled.

Fig. 6.4

6.2   Different Fuels

A second way to make the aviation sector more environmentally friendly is by using different fuels. Instead of using kerosene based on fossil fuels, renewable fuels can be used. For example:

- biofuels made from organic matter and live plants. Biofuels work well with modern jet engines. For example, in 2008, a Boeing 747 flew from London to Amsterdam with one of four engines using a biofuel mixture made from coconut oils and babassu palms.
- hydrogen. Liquid hydrogen can be produced by sending electricity through water so that water molecules (H_2O) are split into oxygen (O_2) and hydrogen (H_2). Electricity needed for this process can be obtained from any renewable energy source. It is a good alternative because engines running on hydrogen emit no CO_2. They also emit less nitrogen oxides (NO_x), which are not greenhouse gases but contribute to creating ozone in the lowest layer of the atmosphere. Ozone is the third most important greenhouse gas after CO_2 and methane. (Higher up in the atmosphere, ozone is helpful, because it protects us from ultraviolet radiation from the sun.)

Using alternative fuels would help reduce the dependence on fossil fuels and reduce greenhouse emissions by as much as 80%.

An alternative to using different fuels is using electric engines. Airplanes with electric engines are currently feasible for small airplanes; whether they are also feasible for large airplanes in the future is unclear yet. For example, the Alice cargo planes are fully electric. In the video in Fig. 6.4, this airplane is explained and compared to a jet-engine airplane:

6.3   Making Airplanes More Efficient

A third way to make the aviation sector more environmentally friendly is by making airplanes more efficient. Airplanes can be made more efficient through new designs and technology, which change the amounts of fuel used.

Fig. 6.5 The Concorde is a tail-less airplane

New designs mean changing airplanes by improving their shapes, structures, and materials. Better designs reduce the weight and drag of the airplane, which improves fuel efficiency. For example:

- using tail-less designs reduces weight and improves aerodynamics. Tailless means that the main wings are the only horizontal aerodynamic parts of the plane (see Fig. 6.5). Aerodynamics involves how air flows past the airplane.
- changing components such as seats, trolleys, paints, and entertainment systems by using lighter materials to reduce the plane's weight.
- adding wingtip extensions to wings improves the airplane's aerodynamics and reduces the amount of fuel needed, even though they add additional weight (see Fig. 6.6). These extensions can also be installed on existing airplanes.

New technology means changing airplanes by improving the technology used in the plane. For example, using new engine designs, such as turbofan engines (see Fig. 6.7), can improve fuel efficiency. These engines can also be installed on existing airplanes.

These changes are especially effective in reducing CO_2 emissions when combined. For example, if the existing Airbus fleet would be replaced by new design airplanes with the latest technologies, the climate impact could be reduced by 1/3, without changing operating costs. As about 7000 Airbuses (A320) are in use, replacing these airplanes or their engines could make air transportation a lot more environmentally friendly.

Fig. 6.6 A wingtip extension

Fig. 6.7 A turbofan engine

6.4 Making Operations More Efficient

A fourth way to make the aviation sector more environmentally friendly is by making operations more efficient. To achieve more efficiency here, new technologies are not always needed. Instead, changes to reduce CO_2 emissions can be made immediately. For example:

- choosing an airplane with the right capacity for each trip will reduce fuel use and emissions. The right capacity especially means not too big.
- routing airplanes flexibly and efficiently using newer air traffic management systems reduces flying times.
- using best practices for ground operations also reduces fuel use. Best practices can, for example, be applied for operations during flight, service and airplane maintenance operations, and auxiliary power units. Auxiliary power units are devices that provide energy for operations other than moving forward, such as turning on the fasten your seat belt lights.
- flying direct routes by improving flight schedules and gate assignments also improves efficiency by up to 11%. And it not only improves fuel efficiency, but it also improves the passenger experience.

6.5 Carbon Offsetting

A fifth way to make the aviation sector more environmentally friendly is by carbon offsetting. Carbon offsetting means investing in a project that reduces the impact of greenhouse gas emissions. An example of a carbon offset project is planting trees. Some airlines have already included the option to carbon offset in their booking procedure. If carbon offsetting programs are expanded internationally, they can reduce the impact of air transportation by 18%.

6.6 Conclusion

So, as the aviation industry puts a lot of pressure on the environment and is at the same time part of modern society, it is important to make the aviation sector more environmentally friendly. This can for example be achieved by replacing kerosene based on fossil fuels with kerosene made from CO_2, water, and sunlight. It is also possible to replace kerosene with other fuels such as biofuels and hydrogen, or to use electric planes.

Apart from replacing fossil fuel-based kerosene, planes can be made more efficient, so that they require less fuel. Also, operations on the ground and in the air can be made more efficient, and flights can be carbon offset. Especially when several of these methods are combined, a large positive difference can be made.

6.7   How We Can Take Action

While the aviation industry can be made more environmentally friendly in many different ways, also we as individuals can make a positive difference. Here are practical ideas of what you and I can do to help reduce the environmental impacts of air travel:

- Using videoconferencing instead of flying
- Traveling by train instead of flying when possible
- Carbon offsetting the flight
- Bringing as little luggage as possible
- Booking direct flights instead of flights with one or more stops
- Asking legislators to regulate climate impacts of air transportation
- Making the flight worth it more by flying somewhere for several reasons, instead of just one reason (for example, going on a business meeting, meeting friends, and a holiday, instead of just going on holiday)
- Making flights worth it more by staying longer, instead of just flying for a long weekend

Credit

This Chapter Is Based On:

Air Travel:
Hasan, M. A. et al. (2021). Climate change mitigation pathways for the aviation sector. *Sustainability*, *13*(7), 3656.

Kerosene:
Zoller, S. et al. (2022). A solar tower fuel plant for the thermochemical production of kerosene from H2O and CO2. *Joule*, *6*(7), 1606–1616.

Figure Credits

Fig. 6.1   Adapted from Jo Sam Re on Shutterstock
Fig. 6.2   Dr. Erlijn van Genuchten
Fig. 6.3   Peter Hermes Furian on Shutterstock
Fig. 6.5   agsaz on Shutterstock
Fig. 6.6   Dr. Erlijn van Genuchten
Fig. 6.7   1968 on Shutterstock

Part II
Pollution

In Volume 1 of A Guide to a Healthier Planet, several types of pollution and environmental consequences were explained: plastic, air, heavy metal, and light pollution:

- Plastic pollution is caused when synthetic plastic products end up in the environment.
- Air pollution is caused when chemical, physical, or biological contaminants in the form of gases, liquid droplets, or tiny solid particles end up in the air.
- Heavy metal pollution is caused when heavy metals such as cadmium, nickel, lead, and manganese end up in the environment.
- Light pollution is caused when human light sources, such as street lights, brighten the night sky. (Further reading: Introduction of Part II of "A Guide to a Healthier Planet" Volume 1).

Unfortunately, these types of pollution have many more consequences. Also, many more types of pollution exist, although we may not always be aware it is considered pollution or we simply don't know it exists! These types of pollution can also directly or indirectly affect our health.

One example of another type of pollution that *directly* affects our health is **noise pollution**. Noise pollution involves harmful or annoying noise levels, for example caused by a busy road (see Fig. 1). Even though this type of pollution is clearly audible and has a significant health impact, we may not always be aware of it.

A second example of pollution that *directly* affects our health is **nitrogen pollution**. Nitrogen (NO_x) coming from different sources, including vehicles that burn fossil fuels and synthetic fertilizers used in agriculture, can cause nitrogen pollution. When fossil fuels are burnt, the air is polluted with nitrogen. While the air in our atmosphere naturally consists of 78% nitrogen gas (N_2), the nitrogen that is considered pollution involves reactive nitrogen atoms. These reactive atoms pair up with other atoms, such as nitrate (NO_3) and nitrogen oxide (NO_x). These molecules contribute to different environmental and health issues.

One example of another type of pollution that *indirectly* affects our health is **soil pollution**. Soil pollution, also called land pollution, is caused when toxic chemicals

Fig. 1 Busy roads often cause noise pollution in surrounding areas

Fig. 2 Oil from a barrel leaking into the ground causes soil pollution

end up in the ground in concentrations high enough to put human health at risk. For example, the soil under landfills is often polluted by substances such as arsenic from chips on electronic devices, lead from batteries, and oil from barrels that leak into the ground (see Fig. 2). From here, it can for example enter our bodies through eating food that has taken up contaminants.

Fig. 3 Sunscreen is important for our health but can be harmful to the environment

A second example of pollution that *indirectly* affects our health is **sunscreen pollution**. Sunscreen is a mixture of oils, water, and ultraviolet filters (see Fig. 3). The ratio between these ingredients depends on the manufacturer's recipe and on the amount of ultraviolet protection, the solar protection factor. There are over 1300 different products, of which more than 850 contain harmful chemicals. These chemicals end up in the water both directly, for example during swimming, or indirectly, for example after having washed the sunscreen off in the shower. As a consequence, areas with more recreational activities, such as swimming, diving, and surfing, have higher levels of harmful chemicals. These harmful chemicals harm marine life, which indirectly harms us as well.

Figure Credits

Chapter 7
How Different Types of Pollution Affect Pollinators

Abstract Environmental pollution has a wide range of consequences, with direct and indirect effects on humans. One of the indirect effects on food security is caused by the negative impact of pollution on pollinators. Pollinator diversity and populations are negatively affected by the use of chemicals in agriculture, and other urban and industrial factors. The chemicals and other factors cause different types of pollution, including air, light, soil, and noise pollution. For example, air pollution negatively affects pollinators by weakening their memory; light pollution affects their reproduction; soil pollution impacts pollen/nectar quality; and noise pollution can increase stress levels in larvae. These and other negative effects contribute to 40% of insect pollinators being at risk of extinction, which in turn puts our food security at risk.

Keywords Science · Science communication · Pollution · Pollution consequences · Pollinators · Biodiversity · Food security · Biodiversity · Conservation · Decline · Ecosystem · Pollutants

The many different types of pollution have a broad range of consequences that are far-reaching. As we often only focus on the direct effects on us, we often don't realize the indirect effects when they affect other organisms such as pollinators. Currently, about 40% of insect pollinators are at risk of extinction. This is a huge problem, as pollinators and plants have evolved together over millions of years, and therefore about 90% of flowering plants depend on pollinators. This means that declining pollinator diversity and populations also indirectly affect us, for example by putting food security at risk. Diversity refers to different pollinator species, including bees, wasps, ants, butterflies, beetles, moths, and hummingbirds.

This decline in pollinator diversity and populations is mostly caused by chemical pesticides, insecticides, and fertilizers. But also other urban and industrial factors types of pollution cause problems. Problems include reduced habitat (see Fig. 7.1) and pollution. This is how different types of pollution affect pollinators:

Credit: This chapter is based on the scientific article "Pollution and pollinators: A review" by Jaweria Riaz and colleagues. (Full citation is available at the end of the chapter)

E. van Genuchten, *A Guide to a Healthier Planet, Volume 2*,
https://doi.org/10.1007/978-3-031-60128-6_7

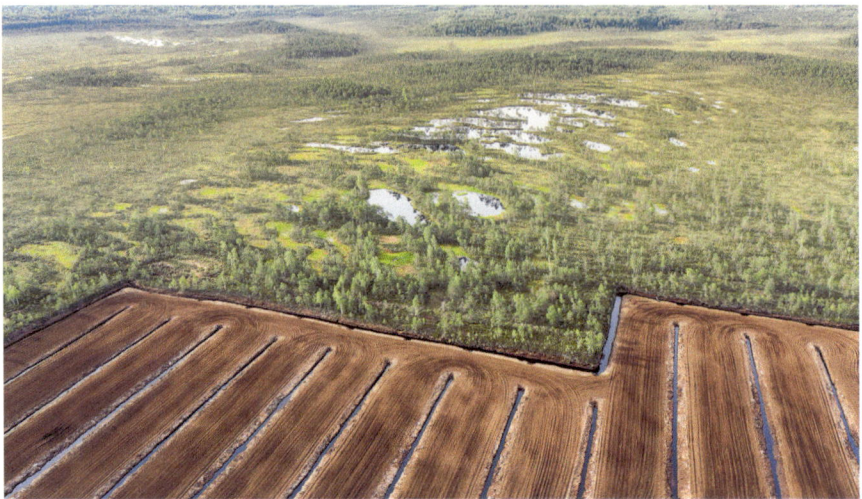

Fig. 7.1 Habitat loss because of human activities contributes to the reduction of pollinators

Fig. 7.2 Air pollution can
cause insects such as
bumblebees to die

7.1   Air Pollution

The first type of pollution that negatively affects pollinators is air pollution. Air pollution is problematic for pollinators because:

- air pollutants can get trapped in leaves, causing the extinction of pollinators that feed on the leaf juices (see Fig. 7.2).
- some pollutants, such as ozone, affect flowers' scent, which they release into the air to attract pollinators. Air pollutants disturb the scent's distribution, which affects how efficiently pollinators can find food.

- the scents released by flowers don't reach several kilometers (1 km = 0.6 mi) anymore, but only a few meters (1 m = 3.3 ft). This means pollinators have to fly longer distances and spend more time searching for food. This weakens them.
- it can affect their memory, their ability to smell, and cognition. This makes it for example harder for the pollinator to recognize plants and find food.

7.2 Light Pollution

The second type of pollution that negatively affects pollinators is light pollution. Light pollution is problematic for pollinators because:

- it disturbs interaction with other species and with other members of the same species. This for example affects reproduction (see Fig. 7.3). (Further reading: Chap. 9 of A Guide to a Healthier Planet Volume 1: "How Light Pollution Impacts Our Environment")
- it affects the number of insects present in an area.
- it reduces their eyesight, as visual cues become less obvious.
- a different light intensity changes when pollinators wake up at night.
- they are more easily caught by predators.
- it can cause competition for proper breeding grounds when insects tend to lay their eggs near light sources.

Fig. 7.3 Light pollution affects the activities of nocturnal insects

Fig. 7.4 Soil pollution can lead to lower habitat quality, for example for this bee

7.3 Soil Pollution

The third type of pollution that negatively affects pollinators is soil pollution. Soil pollution is problematic for pollinators because:

- it can lead to lower habitat quality (see Fig. 7.4).
- it can affect plants, on which pollinators depend for food. For example, plants can become less resistant to pests.
- the quality of pollen/nectar can be affected.
- it affects their interaction with plants. For example, they visit plants shorter, which causes them to gather fewer pollen.
- it can cause memory loss or impair their ability to smell.
- it can impact their gesturing behavior. For example, it affects bees' waggle dance, which is needed to communicate the location of food sources to other bees.
- it can cause chronic toxicity. Chronic toxicity is a negative consequence caused by long-term exposure to pollutants.

7.4 Noise Pollution

The fourth type of pollution that negatively affects pollinators is noise pollution. Noise pollution is problematic for pollinators because it triggers them to become less active and it can increase heart rates in larvae. This shows that noise is stressful for them.

7.5 Conclusion

So, different types of pollution are putting pollinators at risk. These include air, light, soil, and noise pollution – affecting pollinators in different ways. This is critical because pollinator populations have reduced significantly over the last few decades and are further decreasing. Many are even at risk of going extinct! This is not only critically dangerous for them, it is also for humans as they for example significantly contribute to our food security. That is why we must prevent and resolve these types of pollution to help save pollinators.

7.6 How We Can Take Action

To be able to support pollinators and prevent their extinction, it is helpful to reduce these different types of pollution. Here are examples of what you and I can do in daily life to reduce *air pollution*:

- Planting trees and other vegetation that helps filter air pollution (further reading: Chap. 12 of A Guide to a Healthier Planet Volume 1: "Pollution Solutions: Removing Pollutants from Air")
- Using an electric instead of fossil fuel-based vehicle
- Bringing garbage to recycling instead of burning it
- Making sure that old cars have a proper exhaust filter that prevents dirty fumes
- Traveling by bike or train instead of car
- Turning off the engine while at standstill
- Putting special coatings on buildings that remove air pollution (further reading: Chap. 12 of A Guide to a Healthier Planet Volume 1: "Pollution Solutions: Removing Pollutants from Air")

Here are examples of what you and I can do in daily life to reduce *light pollution*:

- Turning off lights when they are not in use
- Keeping blinds and curtains closed when using artificial light inside
- Avoiding driving at night
- Letting eyes adjust to the darkness instead of using light
- Installing fewer outside lights
- Using dark mode of smartphone when using it outside at night
- Using outside lights with motion detection instead of having them on all the time (see Fig. 7.5)

Here are examples of what you and I can do in daily life to reduce *soil pollution*:

- Bringing toxic waste to recycling to reduce soil pollution
- Growing plants that can remove pollution from the soil (further reading: Chap. 11 of A Guide to a Healthier Planet Volume 1: "Pollution Solutions: Removing Pollutants from Soil and Water")

Fig. 7.5 An outdoor motion detector and light

- Using microorganisms to remove pollution from soil (further reading: Chap. 11 of A Guide to a Healthier Planet Volume 1: "Pollution Solutions: Removing Pollutants from Soil and Water")
- Doing a cleanup to remove polluting waste
- Joining an initiative that aims at saving soil

Here are examples of what you and I can do in daily life to reduce *noise pollution*:

- Growing a green fence to keep noise from the road out of the garden
- Closing doors and windows when listening to loud music
- Insulating our home, not only to keep heat but also sound in
- Limiting car use as much as possible
- Being mindful late at night, to not make any disturbing noises
- Using quiet instead of loud vehicles

Credit

This Chapter Is Based On:

Riaz, J. et al. (2020). 04. Pollution and pollinators: A review. *Pure and Applied Biology (PAB)*, *9*(3), 2049–2058.

Figure Credits

Fig. 7.1   F-Focus by Mati Kose on Shutterstock
Fig. 7.2   Paustius on Shutterstock
Fig. 7.3   Fer Gregory on Shutterstock
Fig. 7.4   Wirestock creators on Shutterstock
Fig. 7.5   Bilanol on Shutterstock

Chapter 8
How Sunscreen Pollution Affects Marine Environments

Abstract Sunscreen is essential for protecting skin from harmful ultraviolet sun rays, but its use by millions of people harms the environment. As the number of international tourists worldwide has grown, sunscreen pollution is affecting coastal areas and marine ecosystems in many ways. This is because sunscreen contains chemicals and small particles that block ultraviolet light but can also harm or kill marine organisms. These consequences are far-reaching because this harm does not limit itself to single organisms: it can impact the whole food chain and cause whole coral reef ecosystems to die. That is why it is important to keep protecting our skin but in a more environmentally friendly way.

Keywords Science · Science communication · Pollution · Sunscreen pollution · Pollution consequences · Sunscreens · Marine environment · Coral reef · UV filters · Environment pollution · Nanoparticles · Coastal areas · Ecotoxicology · Nano risk · Ecosystem · Toxic mechanism

When we pack our luggage for a summer holiday, sunscreen is often on the list to be able to protect our skin from too much sun exposure. Protecting our skin is a good idea as too much sun can cause both long-term and short-term health issues. A short-term issue is sunburn, and long-term issues are skin cancer and premature skin aging.

The good news is that many health issues are prevented by using sunscreen. The bad news is that the use of sunscreen by millions of people harms the environment: the number of international tourists worldwide grew from 463 million in 1992 and was expected to be 1.56 billion in 2020 if the COVID-19 pandemic hadn't hit. That is almost 34 times as many tourists! Many of these tourists visit coastal areas and marine environments (see Fig. 8.1). This means that a lot of sunscreen, which contains chemicals, ends up in aquatic environments: in tropical countries, 25% of sunscreen applied to skin ends up in the ocean.

Credit: This chapter is based on four scientific articles by Samuele Caloni, Jérôme Labille, Djordje Vuckovic, and Shengwu Yuan and their colleagues. (Full citations are available at the end of the chapter).

Fig. 8.1 Many of these beach-goers in Rio de Janeiro have used sunscreen for skin protection

The chemicals in sunscreen are mostly used to block ultraviolet light. There are two types of ultraviolet filters: organic and inorganic filters. Some sunscreen products contain both types of filters. This is how organic and inorganic filters damage aquatic environments:

8.1   Organic Filters

The first type of ultraviolet filter in sunscreen are organic filters. Organic filters, also called chemical filters, are made up of chemical compounds. These chemical compounds are complex molecules made of natural materials combined with chemicals. These complex molecules release energy as fluorescence or heat when hit by ultraviolet light. This energy neutralizes the ultraviolet rays. Unfortunately, they also affect aquatic environments by harming the animals in them, for example:

- They kill zooxanthellae that live on coral reefs, which causes the reef to bleach. These one-celled algae live in a symbiotic relationship with corals (see Fig. 8.2). A symbiotic relationship is a close relationship between organisms of different species that are harmless to or even benefit each other. When these zooxanthellae die, the corals die as well. This can lead to the death of the whole coral reef and in turn a loss of biodiversity.

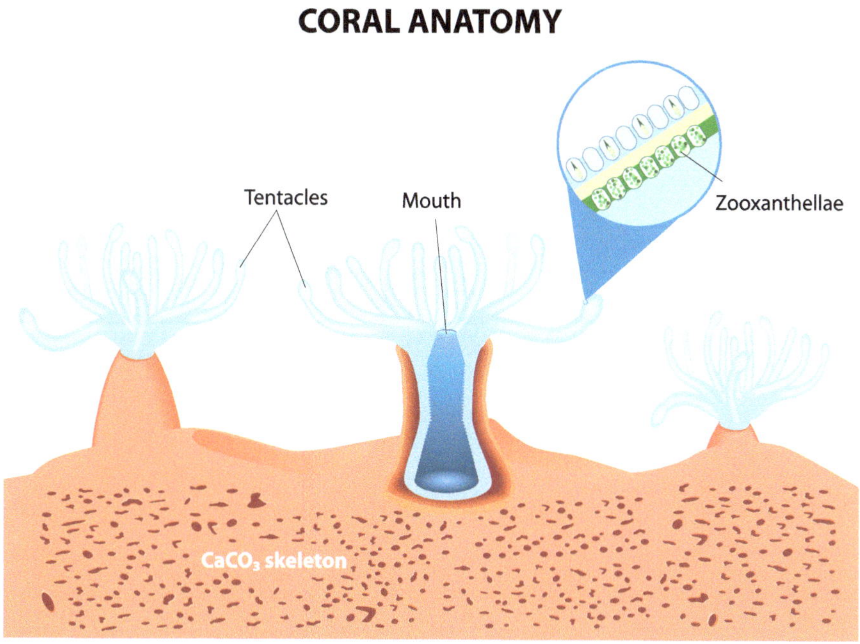

CORAL ANATOMY

Fig. 8.2 Zooxanthellae live on and in a symbiotic relationship with corals

- They damage corals and anemones when the chemicals pair up with sugar molecules. As a pair, they cause the coral and anemone surfaces to oxidize when light hits them. Oxidation is a chemical reaction that removes an electron from a molecule, which harms the tissue. When the number of these chemical-sugar pairs grows, corals and anemones are more likely to die. This negative effect is even larger when algae that live in a symbiotic relationship with these animals are lacking, as they can reduce the number of these harmful pairs.
- They cause hormonal changes in different marine animals. These changes affect the function of many body systems. For example, the rainbow trout's (see Fig. 8.3) reproductive and respiratory systems are affected by certain chemicals found in sunscreen. And in fathead minnows, these chemicals reduce fertility, change reproductive cells, and even cause male minnows to become more female.
- They cause an overgrowth of phytoplankton. Phytoplankton are an early part of the food chain, which means they are food for many other animals that live in the aquatic environment. Changes in the number of phytoplankton can change the structure of the food chain.
- They can cause algae blooms, as these chemicals also support algae growth. Such blooms cause a lack of oxygen in the water, which kills other animals (see Fig. 8.4).

Fig. 8.3 Illustration of a rainbow trout fish

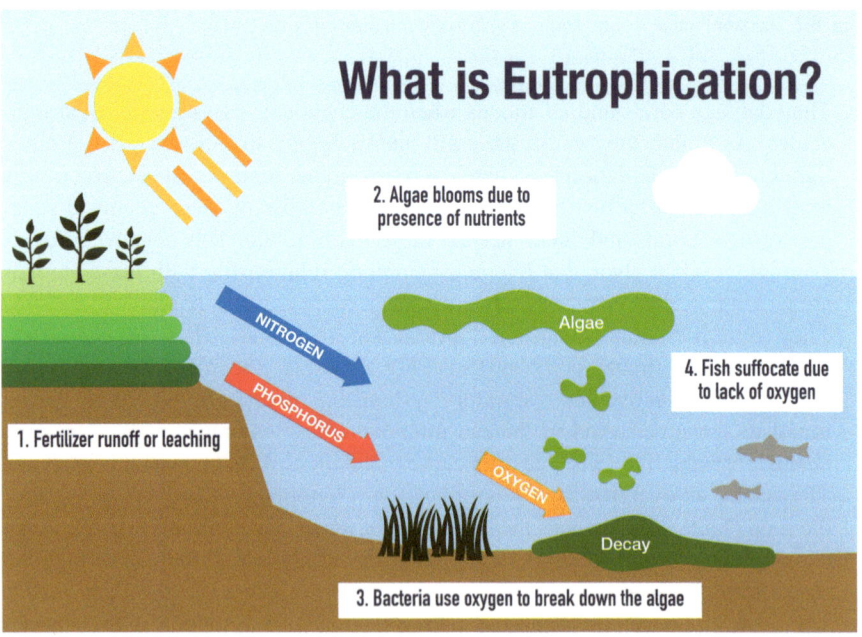

Fig. 8.4 Algae blooms, in this image just under the water surface, can cause fish to suffocate

8.2   Inorganic Filters

The second type of ultraviolet filter in sunscreen are inorganic filters. Inorganic filters, also known as physical filters or mineral filters, contain metal nanoparticles. Nanoparticles are very small particles of between 1 and 100 nanometers (nm). To compare, a hair is between 80,000 and 100,000 nm, or up to 1000 times as wide. Titanium dioxide or zinc oxide are most commonly used. These nanoparticles reflect and scatter ultraviolet rays.

Inorganic filters have been produced as an alternative to harmful organic filters. But unfortunately, these filters are harmful to the environment and animals living in it as well. This is because sunscreens with inorganic filters increase the levels of high-risk, heavy metals, such as aluminum, zinc, magnesium, copper, lead, and chromium in aquatic environments.

While the level of high-risk metals increases in the water, it is also more and more often found in sediment and animals. Once they are in the environment, they stay in the environment for a long time, potentially harming many types of aquatic animals. For example:

- Titanium dioxide causes certain algae to grow more slowly and higher concentrations can damage their cells.
- Certain phytoplanktons' cells and maybe even their genes are damaged by nanotitanium dioxide.
- Sea urchins exposed to nanotitanium dioxide stop developing normally: their growth slows down and they can develop deformities in their skeletons. This is because they have lower levels of certain enzymes, which speed up chemical reactions.
- Stony coral and certain shrimps die when being exposed to zinc oxide.

And they can harm us as well. For example, heavy metals can contribute to Parkinson's disease (further reading: Chap. 10 of A Guide to a Healthier Planet Volume 1: "How Heavy Metal Pollution Can Cause Parkinson Disease").

8.3   Conclusion

So, while protecting our skin from the sun is important for our health, the resulting sunscreen pollution in for example marine environments is critical. This is because the chemicals and particles in sunscreen harm marine life in many different ways. For example, chemicals kill algae that protect corals, causing them to bleach and die, and cause hormonal changes in certain fish, and changes in the food chain. They can even indirectly cause fish to suffocate. Also, particles can damage animals' cells, hinder marine animals' normal development, or even cause them to die. That is why it is important to keep protecting our skin but in a more environmentally friendly way.

Fig. 8.5 Sun protection clothes instead of sun screen can be used during water activities to prevent sunscreen pollution

8.4   How We Can Take Action

As sunscreen pollution is so harmful to marine environments, it is important to limit the amount of sunscreen that ends up in waterways. Here are practical ideas of what you and I can do to reduce sunscreen pollution in daily life:

- Using protective clothing instead of sunscreen when possible (see Fig. 8.5)
- Using sunscreen products carefully, applying only the amount needed
- Using sunscreen after water activities instead of before when possible
- Using sunscreen products that are proven to be environmentally friendly
- Verifying that your local waste treatment facilities are equipped to handle heavy metals from sunscreen

Credit

This Chapter Is Based On:

Caloni, S., Durazzano, T., Franci, G., & Marsili, L. (2021). Sunscreens' UV filters risk for coastal marine environment biodiversity: A review. *Diversity, 13*(8), 374.

Labille, J., Catalano, R., Slomberg, D., Motellier, S., Pinsino, A., Hennebert, P., … & Bartolomei, V. (2020). Assessing sunscreen lifecycle to minimize environmental risk posed by nanoparticulate UV-filters–A review for safer-by-design products. *Frontiers in Environmental Science, 8*, 101.

Vuckovic, D., Tinoco, A. I., Ling, L., Renicke, C., Pringle, J. R., & Mitch, W. A. (2022). Conversion of oxybenzone sunscreen to phototoxic glucoside conjugates by sea anemones and corals. *Science, 376*(6593), 644–648.

Yuan, S., Huang, J., Jiang, X., Huang, Y., Zhu, X., & Cai, Z. (2022). Environmental fate and toxicity of sunscreen-derived inorganic ultraviolet filters in aquatic environments: A review. *Nanomaterials, 12*(4), 699.

Figure Credits

Fig. 8.1   R.M. Nunes on Shutterstock
Fig. 8.2   Based on Designua on Shutterstock
Fig. 8.3   Moloko88 on Shutterstock
Fig. 8.4   Dimitrios Karamitros on Shutterstock
Fig. 8.5   Anna Kuzmenko on Shutterstock

Chapter 9
How Nitrogen Pollution Affects Pollen Allergies

Abstract Nitrogen pollution caused by, for example, burning fossil fuels can increase the presence of pollen allergies. While some have noticed an increase in their allergic reactions in recent years, many are unaware that this may be caused by nitrogen pollution. This type of pollution impacts us through various mechanisms. These mechanisms impact the number of pollen in the air, the plants that release pollen, the pollen themselves, or our reaction to pollen. As a consequence, we react differently and/or more strongly, leading to more and worse symptoms.

Keywords Science · Science communication · Pollution · Nitrogen pollution · Nitrogen · Pollution consequences · Planetary health · Allergic disease · Environmental pollution · Pollen

Apart from consequences of environmental pollution going unnoticed, it is also possible that consequences are noticed but that the cause remains unclear. For example, nitrogen pollution caused by burning fossil fuels can impact pollen allergies, but while some have noticed an increase in their allergic reactions in recent years (see Fig. 9.1), many are unaware that this may be caused by nitrogen pollution.

This impact of nitrogen pollution on pollen allergies is caused by various mechanisms. This is how nitrogen pollution can increase the presence of pollen allergies:

9.1 Changing Pollen Exposure

The first way nitrogen pollution can increase the presence of pollen allergies is by changing pollen exposure. Pollen exposure can be changed, as more reactive nitrogen molecules surround us when the air is polluted by more nitrogen. These reactive nitrogen molecules change plant communities, so that some plants become more

---

Credit: This chapter is based on the scientific article "Impact of environmental nitrogen pollution on pollen allergy: A scoping review" by Paulien Verscheure and colleagues. (Full citation is available at the end of the chapter)

ALLERGY SYMPTOMS

Fig. 9.1 A pollen allergy can cause different symptoms

Fig. 9.2 Pollen have different structures, which can be changed by nitrogen pollution

and others less dominant. In turn, with different plant communities, also pollen exposure changes. Depending on the changes, this can lead to more allergic reactions.

9.2 Affecting Pollen Shape and Structure

The second way nitrogen pollution can increase the presence of pollen allergies is by affecting the pollen membrane and morphology (see Fig. 9.2). The membrane is the outer layer of the pollen. This means that nitrogen pollution can change the shape and structure of pollen.

The shape and structure of a pollen changes when nitrogen is absorbed or has damaged the membrane. As a consequence, the pollen become smaller and more fragile. Also, the membrane becomes thinner so that it breaks more easily. The resulting differences in the pollen shape and structure can lead to different allergic reactions in our bodies.

9.3 Changing Pollen Proteins

The third way nitrogen pollution can increase the presence of pollen allergies is by changing pollen proteins. Proteins in pollen are changed because chemical modifications are triggered by nitrogen pollution. For example, nitric oxide (NO) groups are incorporated into the proteins. A possible consequence of chemical modifications is that the protein structure becomes looser so that they become more stable and stay in the air for a longer time.

9.4 Impacting Allergen Release

The fourth way nitrogen pollution can increase the presence of pollen allergies is by impacting allergen release. Allergens are substances that cause allergic reactions. For example, sub-pollen particles are allergens. Sub-pollen particles are tiny pollen particles that are so small that they can reach the lower airways, the smallest parts of the lungs. When a pollen membrane is damaged by nitrogen pollution, more of these sub-pollen particles can be released and our lungs can be affected more.

9.5 Affecting Fertilization

The fifth way nitrogen pollution can increase the presence of pollen allergies is by affecting how likely pollen can fertilize plants. In general, nitrogen pollution makes pollen less likely to successfully fertilize a plant. But this also depends on the species: some plants are less likely to germinate when being affected by nitrogen pollution, other plants are even more likely to grow, and for other plants it doesn't make a difference. While it is unclear whether these differences in germination matter for allergic reactions, it is clear that affected fertilization can go together with releasing more allergens.

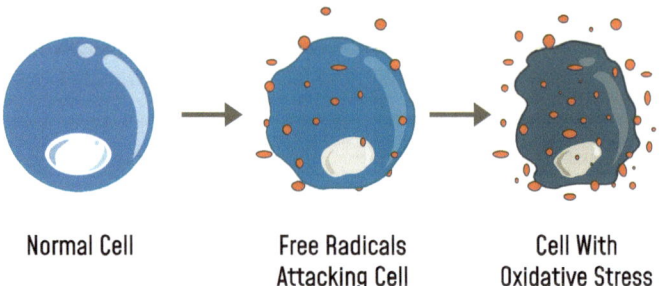

Fig. 9.3 Oxidative stress can affect pollen

9.6 Triggering Oxidative Defense Mechanisms

The sixth way nitrogen pollution can increase the presence of pollen allergies is by changing the oxidative defense mechanism. The oxidative defense mechanism is a mechanism that protects a plant against oxidative stress (see Fig. 9.3). Oxidative stress is caused when highly reactive molecules are present in the plant. Nitrogen pollution activates the oxidative defense mechanism, which causes chemical modifications, a different protein structure of pollen, or both. This in turn makes allergic reactions more likely.

9.7 Increasing Allergenic Reactivity

The seventh way nitrogen pollution can increase the presence of pollen allergies is by increasing allergenic reactivity. Increased allergenic reactivity means that humans' immune systems respond more strongly to a specific allergen. When pollen are affected by nitrogen pollution, our bodies react stronger. Also, breathing air with nitrogen pollution can make our airways more sensitive. This can also increase our reactivity to pollen particles.

9.8 Worsening Symptoms

The eighth way nitrogen pollution can increase the presence of pollen allergies is by worsening allergic symptoms. For example, allergic symptoms in eyes and nose become stronger in allergic people when they are exposed to nitrogen pollution.

This can be recognized not only by bodily reactions but also by more sales of anti-histamines. Antihistamines are drugs that inhibit the body's reaction to allergens.

9.9 Conclusion

So, while nitrogen pollution impacts pollen allergies and the consequences are noticeable through allergic reactions, the link to environmental pollution often goes unnoticed. This may be because pollen particles are tiny, some even so small that they can reach the smallest parts of our lungs.

The impact of nitrogen pollution on pollen allergies is caused by various mechanisms. These mechanisms impact the number of pollen in the air, the plants that release pollen, the pollen themselves, or our reaction to pollen. Especially increased allergic reactions and worsened symptoms can help us recognize the negative impact of nitrogen pollution on our health.

9.10 How We Can Take Action

As nitrogen pollution can impact our health through worsening pollen allergies, it is important to limit nitrogen pollution. Here are practical ideas of what you and I can do to reduce nitrogen pollution:

- Enjoying a virtual walk instead of traveling around the globe (see Fig. 9.4)
- Traveling by public transport instead of driving a car
- Carpooling instead of traveling in separate cars
- Reducing meat and dairy intake
- Buying renewable energy from energy provider instead of energy based on fossil fuels
- Driving an electric or hydrogen car instead of gasoline or diesel car
- Discharging wastewater from toilets appropriately so that it doesn't leak into soil and groundwater
- Limiting fertilizer use as much as possible

Fig. 9.4

Credit

This Chapter Is Based On:

Verscheure, P., et al. (2023). Impact of environmental nitrogen pollution on pollen allergy: A scoping review. *Science of the Total Environment,* 164801.

Figure Credits

Fig. 9.1   PCH.Vector on Shutterstock
Fig. 9.2   nobeastsofierce on Shutterstock
Fig. 9.3   Fancy Tapis on Shutterstock

Chapter 10
How Plastic Pollution Can Spread Viruses

Abstract Plastic pollution is a growing concern because it can impact our health. This can happen when tiny plastic particles enter our bodies, but also when plastic pollution spreads harmful viruses. Whether plastic can spread viruses depends on viral, plastic, and environmental characteristics. Viral characteristics include whether viruses clump together on the plastic surface, whether they have an envelope, and the surface charge. Plastic characteristics include roughness, hydrophobicity, and surface charge. Environmental characteristics include temperature, water salinity, and the amount of organic matter dissolved in the water. Interactions between viral, plastic, and environmental characteristics also play a role, as viruses with different characteristics can react differently to environmental conditions.

Keywords Science · Science communication · Pollution · Plastic pollution · Pollution consequences · Microplastics · Virology · Health consequences · Biofilm · Environmental virology · Wastewater

While nitrogen pollution can impact our health by affecting pollen allergies, this is just one of the many ways in which environmental pollution affects our health. Another example is plastic pollution, which can impact our health after we have inhaled or taken in plastic particles. Inhaling plastic particles is possible because these particles are so small that they can easily be transported. Also; these tiny plastic particles can enter the food chain because they are accidentally eaten by marine animals, which are eaten by fish that end up on our plates (further reading: Chap. 8 of A Guide to a Healthier Planet Volume 1: "How Plastic Pollution Impacts Aquatic Animals").

Apart from entering our bodies, plastic pollution can also in a completely different way impact our health: by spreading viruses. Plastics in surface water (see

Credit: This chapter is based on the scientific article "Survival of human enteric and respiratory viruses on plastics in soil, freshwater, and marine environments" by Vanessa Moresco, David M. Oliver, Manfred Weidmann, Sabine Matallana-Surget, and Richard S. Quilliam. (Full citation is available at the end of the chapter)

Fig. 10.1 Microplastics floating around in water can carry viruses

Fig. 10.1) can spread viruses without proper plastic waste management because microorganisms quickly start growing on the plastics' surfaces. Together they can form a very thin layer that separates the environment and the plastic surface. This thin layer is called the plastisphere.

The ability of plastic to spread viruses is concerning because plastic pollution is rapidly increasing. Every year, up to 10^{15} and 10^{16} microplastics enter the environment! That is 125,000 to 1.25 million times the number of people on this planet! It is also concerning because microplastics are hard to filter out of wastewater: after treatment, sewage sludge can still contain 60–99% of the microplastics that entered the wastewater treatment plant!

How well viruses survive on the surface of plastic and therefore how likely plastic pollution spreads viral diseases depends on several factors. These are factors that influence how dangerous plastic pollution can become as a virus spreader:

10.1   Viral Characteristics

The first factor influencing how well plastic pollution can carry viruses is viral characteristics.

One viral characteristic is whether viruses clump together on the plastic surface. When clumping together, they are more stable and degrade less quickly when exposed to heat or ultraviolet radiation from the sun. It also makes them more resistant to saltwater. This increases resistance and makes them more likely to survive.

Enveloped vs non-enveloped virus

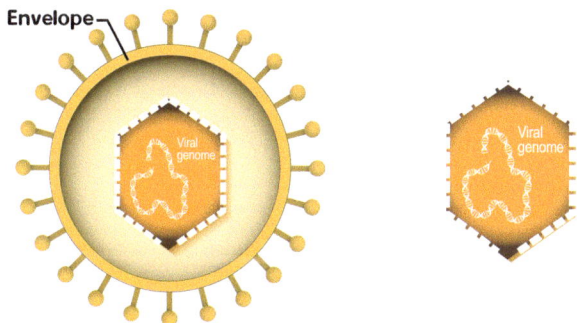

Fig. 10.2 Difference between a virus with an envelope (*left*) and a virus without an envelope (*right*)

Another viral characteristic is whether a virus has an envelope or not (see Fig. 10.2). An envelope is the outer wrapping of a virus. It is not clear how and how well enveloped viruses survive in the environment, but as they are responsible for a large number of infections worldwide, they likely persist well on certain surfaces, including plastic surfaces. But they are less resistant than non-enveloped viruses to wastewater treatment because detergents and solvents damage the envelope.

10.2   Plastic Characteristics

The second factor influencing how well plastic pollution can carry viruses is plastic characteristics.

One plastic characteristic is whether the surface is rough or smooth. Surfaces can be rough for example because of void spaces in the material or small cracks caused by aging. Aging means degradation by sunlight or microorganisms, or breaking. These void spaces give viruses a chance to enter the plastic and gather inside. Small cracks cause the surface to be larger so that more viruses can attach.

Another plastic characteristic is how hydrophobic the surface is. Hydrophobicity means that water is repelled. With higher hydrophobicity, viruses can attach to the surface more easily. This is because viruses are also hydrophobic, which makes it easier to interact.

A third plastic characteristic is the surface charge. When the surface of microplastic particles is charged, electrostatic interaction (see Fig. 10.3) with viruses is possible. Electrostatic interaction means attraction or repulsion between charged molecules. Viruses are naturally negatively charged.

Fig. 10.3 Example of an electrostatic interaction between hair and a balloon

10.3   Environmental Characteristics

The third factor influencing how well plastic pollution can carry viruses is environmental characteristics.

The most important environmental characteristic is temperature. Temperature can either directly influence the structure of viruses on the plastic surface or facilitate processes that can influence their stability. For example, some viruses can be degraded with temperatures over 25 °C (77 °F), but others only over 50 °C (122 °F).

Another environmental characteristic is water salinity. Sometimes higher salt concentrations harm viruses, for example when they become less stable in salty water. But in other cases, higher salt concentrations can help viruses, for example when salt triggers bacteria to secrete a certain type of substance. When this substance clumps together, viruses are more likely to stick and survive.

A third environmental characteristic is how much organic matter is dissolved in the water. Dissolved organic matter is for example plant leftovers and soluble particles that are released by organisms such as bacteria, plants, and algae. When less organic matter is dissolved, non-enveloped viruses are more likely to interact with the plastic surface. This is because hydrophobic forces that are needed for viruses to stick are stronger and because viruses are less likely to be electrostatically repelled by the plastic surface.

Fig. 10.4 The difference between DNA and RNA is that DNA stores and transfers genetic information and RNA is a messenger

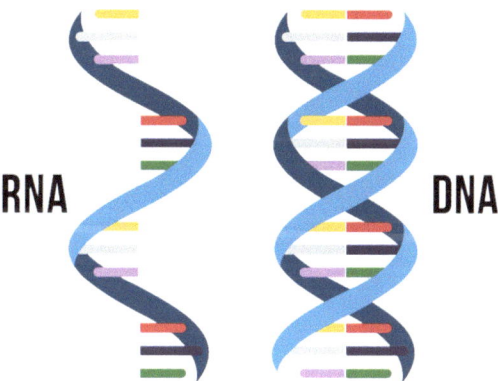

10.4   Interactions

Apart from viral, plastic, and environmental characteristics influencing whether plastic can carry viruses, also interactions between these characteristics are relevant. For example, whether a virus contains DNA or RNA can make them react differently to environmental circumstances. The difference between DNA and RNA is that DNA stores and transfers genetic information, whereas RNA is a messenger between DNA and the cells that produce proteins (see Fig. 10.4). These different reactions influence whether they survive and/or how infectious they are.

10.5   Conclusion

So, while plastic pollution can directly impact our health, it can also indirectly impact our health by spreading viruses. How likely plastic pollution spreads viral diseases depends on many factors. These factors include viral characteristics, such as whether viruses clump together and whether they have an envelope or not. It also depends on plastic characteristics, such as the roughness of the surface, whether the surface is hydrophobic, and whether the surface is charged. It also depends on environmental characteristics, such as temperature, the salinity of water, and whether organic matter is dissolved in the water. And interactions between these characteristics further influence whether viruses survive and/or how infectious they are.

Fig. 10.5 Confetti made from old leaves is biodegradable and more environmentally friendly

10.6   How We Can Take Action

As plastic pollution on surface water can spread viruses and therefore affect our health, it is important to limit plastic pollution as much as possible. Here are practical ideas of what you and I can do to prevent viral infections from spreading on plastic waste:

- Preventing littering to reduce plastic pollution
- Removing plastic from nature that has already been littered
- Recycling plastic instead of sending it to landfill
- Bringing reusable bag to shop to prevent using a new plastic bag
- Buying products without plastic wrapping
- Refraining from letting a balloon up
- Using reusable instead of single-use bottles
- Making confetti from leaves found on the ground instead of buying plastic confetti (see Fig. 10.5)

Credit

This Chapter Is Based On:

Moresco, V., Oliver, D. M., Weidmann, M., Matallana-Surget, S., & Quilliam, R. S. (2021). Survival of human enteric and respiratory viruses on plastics in soil, freshwater, and marine environments. *Environmental Research, 199*, 111367.

Figure Credits

Fig. 10.1   xalien on Shutterstock
Fig. 10.2   Adapted from Designua on Shutterstock
Fig. 10.3   Westock Productions on Shutterstock
Fig. 10.4   Adapted from freaktor on Shutterstock
Fig. 10.5   Maramorosz on Shutterstock

Chapter 11
Pollution Solutions: Removing Plastic Waste from the Environment

Abstract Plastic pollution is a major environmental issue, causing harm to aquatic animals and the environment. To prevent plastic from reaching our oceans, various techniques have been developed, including filtering storm- and wastewater, preventing littering, removing microplastics from wastewater, and preventing microplastics from entering wastewater. But as already a lot of plastic has reached our waterways, it is also important to remove plastic from rivers and oceans. Effective techniques include large-scale booms, river booms, waterway litter traps, robots, boats, wheels, and beach cleaners. These techniques help prevent plastic from causing long-term harm.

Keywords Science · Science communication · Pollution · Plastic pollution · Pollution solutions · Cleanup techniques · Ocean cleanup · Plastic pollution cleanup · Plastic removal methods · Projects cleaning oceans · Ocean cleaners · Inventions for ocean cleanup

As plastic pollution is one of the major types of pollution plaguing our planet, it is important to prevent plastic from reaching the environment. This is because "preventing is better than healing" as a well-known Dutch saying says: when plastic never ends up in the environment, it can't do any harm there.

But while "preventing is better than healing", already a huge amount of plastic pollution is littered in our environment. And this amount keeps growing. This is critical, as plastic that seems to be gone when it is eaten by animals, will return once these animals die and degrade.

For example, plastic in rivers and oceans already causes a lot of harm to aquatic animals (further reading: Chap. 8 of A Guide to a Healthier Planet Volume 1: "How Plastic Pollution Impacts Aquatic Animals"). And shockingly, it is expected that by 2050, the amount of plastic in our oceans will outweigh fish! So, to prevent plastics

---

Credit: This chapter is based on the scientific article "Plastic pollution solutions: emerging technologies to prevent and collect marine plastic pollution" by Emma Schmaltz and colleagues. (Full citation is available at the end of the chapter)

from doing long-term harm to our environment and animals, it is important to remove plastics from our environment as well.

To remove plastic pollution from our environment, different techniques have been developed. Some of these techniques prevent plastics from reaching rivers and oceans. This is effective as a lot of plastic is washed into rivers, which carry plastic over long distances and often eventually bring it to the sea or ocean. Other techniques can remove plastics that are already floating around from rivers and oceans. This is how this can be done:

11.1 Preventing Plastics from Reaching Rivers and Oceans

11.1.1 Preventing Littering

The first way to prevent plastic from reaching rivers and oceans is by preventing littering. Even when waste is left on the ground, rain might wash it into the drainage system.

An example project to prevent littering is "Stow it, Don't Throw it" (see Fig. 11.1). In this project, tennis ball containers are repurposed into fishing line bins for anglers. These bins make it easier to carry these lines to larger recycling bins. This is important, as fishing lines are often eaten by animals or they entangle them, causing them to suffer or even die.

Another example is Goby the fish (see Fig. 11.2). This is a large structure on a beach in Vietnam, that allows visitors to dispose of their plastic waste. Apart from functioning as a bin, it also raises awareness about the importance of preventing plastic from entering the ocean. Similar fish have been installed in different countries.

Fig. 11.1

Fig. 11.2

Fig. 11.3

11.1.2 Preventing Microplastic from Entering Wastewater

The second way to prevent plastic from reaching rivers and oceans is by preventing microplastics from entering wastewater. As an important source of microplastics is microfibers from clothing, laundry balls are helpful. These balls capture microplastics that are set free during washing and laundry.

An example is the Cora Ball (see Fig. 11.3). This ball is put in the washing machine together with clothes. Afterward, microfibers can be removed and disposed of in the bin.

Another example is the Lint LUV-R (see Fig. 11.4). It is similar to the "socks" used to filter storm and wastewater. It is made from stainless steel and captures microfibers when water leaves the washing machine.

A third example is the Guppyfriend Washing Bag (see Fig. 11.5). This bag captures microplastics that have come off the clothes inside the bag during washing. This is highly effective in preventing thousands of tiny fibers from entering wastewater.

11.1.3 Filtering Storm and Wastewater

The third way to prevent plastic from reaching rivers and oceans is by using storm-water and wastewater filters. Such filters are attached to the drainage system that deals with waste or stormwater.

Fig. 11.4

Fig. 11.5

Fig. 11.6 Three StormX netting trash traps

An example is the StormX Netting Trash Trap (see Fig. 11.6). This trap looks like a large sock that is put around the exit of a drainage pipe. Plastic stays behind in the sock, while water flows through it. This prevents any trash larger than 5 mm (0.2 in) from entering rivers.

11.1.4 Removing Microplastics from Wastewater

The fourth way to prevent plastic from reaching rivers and oceans is by removing microplastics from wastewater. This can be done during wastewater treatment.

An example project that investigated how this can be done is the Go Jelly Project (see Fig. 11.7). They use the substance that is secreted by jellyfish when they are stressed to capture microplastics.

11.2 Removing Plastic from Rivers and Oceans

11.2.1 Large-Scale Booms

The first way to remove plastic from rivers and oceans is by using large-scale booms, which are large floating pipes. They capture trash by stopping floating rubbish. This method makes it possible to collect large amounts of plastic, without harming the underneath marine life.

For example, The Ocean Cleanup project (see Fig. 11.8) uses large-scale booms to clean up trash in the Great Pacific Garbage Patch. This boom is C-shaped and floats on the ocean surface in the direction of the current. Floating trash gets trapped, is collected, and then transported to plants for processing.

Fig. 11.7

Fig. 11.8

11.2.2 *River Booms*

The second way to remove plastic from rivers and oceans is by using river booms (see Fig. 11.9). River booms are similar to large-scale booms as they also stop trash. They are usually smaller, extending from one bank to the other side of the river.

For example, the Plastic Fischer Trash Boom (see Fig. 11.10). It consists of PVC pipes that make the steel grid below the surface float. The advantage of these grids is that plastics up to 60 cm deep (23.6 in) can be captured.

Fig. 11.9 A river boom

Fig. 11.10

Fig. 11.11 An Ocean Cleanup interceptor

11.2.3 Waterway Litter Traps

The third way to remove plastic from rivers and oceans is by using litter traps. Litter traps also use booms, but are not only barriers, they also guide waste to a litter trap. This makes taking the plastic out of the river easier.

For example, The Ocean Cleanup interceptor (see Fig. 11.11) is designed to capture waste in flowing streams. A boom guides waste into the interceptor.

11.2.4 Robots

The fourth way to remove plastic from rivers and oceans is by using robots. Robots combined with artificial intelligence can be used to pick up floating plastics. They work very well in water without current. As they are also able to detect fish, they can make sure to only capture plastic.

For example, FRED is a "Floating Robot for Eliminating Debris" (see Fig. 11.12). It has two solar cells that provide power to the collecting vents, which capture floating (plastic-)waste.

Another example is WasteShark (see Fig. 11.13). It has an open "mouth" that vacuums microplastics, biomass, and any other floating waste in rivers and canals.

Fig. 11.12

Fig. 11.13

11.2.5 Boats

The fifth way to remove plastic from rivers and oceans is by using boats. Boats use different tools such as nets to collect all kinds of plastic trash. The advantage is that they can work on a larger scale than for example robots.

For example, ERVIS is a saucer-like ship, that captures waste of any size (see Fig. 11.12). It even analyzes the waste and separates it into five categories, based on the size and type of waste. This makes it easy to dispose of the collected waste correctly (Fig. 11.14).

11.2.6 Wheels

The sixth way to remove plastic from rivers and oceans is by using wheels. Wheels can be put in suitable places, so they can capture waste and transport it with a conveyor belt in a container. The advantage is that they can work day and night.

For example, Mr. Trash Wheel (see Fig. 11.15) is stationed in places where a lot of floating plastic passes. Large water wheels are turned by the flowing water and move the conveyor belts. The captured trash gets trapped in the steps of the conveyor belt. At the end of the belt is a container. When the container is full, the waste is disposed of correctly.

Fig. 11.14

Fig. 11.15 Mr. Trash wheel

11.2.7   Beach Cleaners

The seventh way to remove plastic from rivers and oceans is by using beach clean-ers. This method helps remove plastic that has ended up on beaches. The advantage is that it prevents plastics from being carried back into the ocean and cleans the beach.

For example, the Barber Sand Man is like a hoover that sifts the sand (see Fig. 11.16).

Another example is the Microplastic Buoyancy Filtration Device (see Fig. 11.17), which uses water to filter out the tiny plastic from the sand. The lightest pieces of plastic that float are transported by the water into a container.

Fig. 11.16

Fig. 11.17

Fig. 11.18

11.2.8   Other Methods

Other methods have been invented as well. They need to be researched further to make sure they can be used on a large scale as well.

For example, a technique has been invented to remove microplastics by using a magnet (see Fig. 11.5). To be able to use a magnet, a magnetic powder is mixed with oil, which binds to the microplastics (Fig. 11.18).

11.3   Conclusion

So, as plastic pollution is harming the environment and animals, it is important to prevent further plastic waste from entering the environment. Including waterways such as rivers and oceans. And it is important to remove plastic waste that is already floating around, as this is already causing harm and will be able to do so for a long time.

Apart from preventing plastic from being litter in the first place, it is helpful to use techniques to remove plastic from the environment. As a lot of waste is washed into streams and rivers, and eventually reaches oceans and seas, some techniques focus on preventing litter from reaching waterways. These techniques involve filtering storm and wastewater, preventing littering, removing microplastics from wastewater, and preventing microplastic from entering wastewater.

Other techniques focus on removing litter from waterways. These techniques involve large-scale booms, river booms, waterway litter traps, robots, boats, wheels, beach cleaners, and other methods.

11.4 How We Can Take Action

As plastic pollution harms the environment and organisms, it is important to limit plastic pollution as much as possible. Here are practical ideas of what you and I can do to reduce plastic pollution:

- Taking plastic waste home instead of leaving it behind in the environment
- Picking up waste and putting it in a bin
- Joining a cleanup group to go litter-picking on a regular basis
- Organizing a cleanup event
- Informing your city when you come across a large pile of waste
- Using a product in the washing machine that prevents tiny textile particles from entering wastewater

Credit

This Chapter Is Based On:

Schmaltz, E. et al. (2020). Plastic pollution solutions: Emerging technologies to prevent and collect marine plastic pollution. *Environment International*, *144*, 106067.

Figure Credits

Chapter 12
Pollution Solutions: Removing Waste from Landfills

Abstract Landfills are a significant environmental concern, as they cause environmental pollution, contribute to climate change, and use a lot of land. As landfills grow in size and number, the idea of recovering resources from old landfills has become popular called landfill mining. Landfill mining requires materials to be separated first, which can be achieved using mechanical methods such as milling, sieving, and crushing, and using artificial intelligence. Using artificial intelligence requires training the system to recognize objects and materials. After separation, materials can be put to good use. For example, glass waste can be used for concrete, bricks, tiles, foam glass, water filtration, and sandblasting. Also, landfills can be used to generate electricity, either by capturing and using gases produced by decaying materials or by burning materials that cannot be recycled.

Keywords Science · Science communication · Pollution · Landfills · Pollution solutions · Waste management · Recycling · Landfill mining · Waste glass · Deep learning · Sustainable self-recycling · Neural networks · Transfer learning · Data augmentation · Waste characterization · Municipal solid waste · Biochemical methane potential · Excavated waste

Apart from removing waste from the environment, it is helpful to remove waste from landfills. This is because landfills have been a concern for a long time, as they also harm the environment. For example, heavy metals can leak into the ground, eventually reaching groundwater, and landfills cause a lot of methane emissions, which contribute to climate change (further reading: Chap. 6 of A Guide to a Healthier Planet Volume 1: "Climate Solutions: Controlling Methane Levels"). Also, landfills use land that could be used more effectively, for example, as habitat for wildlife, which supports biodiversity.

Credit: This chapter is based on three scientific articles by Isabella Pecorini, Sergio Náñez Alonso, and Danish Kazmi and their colleagues. (Full citations are available at the end of the chapter)

Apart from causing environmental issues, landfills also cause resources to become scarcer while resources that are piled up remain unused. This is unfortunate as these resources can be put to good use again. Luckily, with landfills growing both in size and number and resources becoming scarcer, the idea of recovering resources from old landfills has become popular. This means that landfills are turned into temporary storage instead of a final destination.

Recovering resources is called landfill mining and has three benefits: the negative impact on the environment is reduced, more land can be put to good use, and excavated waste can be used for cost-effective recovery of resources. Here are examples of resources that can be mined from landfills, turning excavated waste into tomorrow's resource:

12.1   Removing Recyclables

The first way to turn excavated waste into tomorrow's resource is by recovering reusable materials. Old landfill waste looks like home recycling, containing a lot of reusable materials, such as glass, paper, plastic, wood, and metals. All of these materials have value and need to be sorted and separated first (see Fig. 12.1).

But sorting and separating excavated waste is not a trivial task. If done by hand, this processing is slow and dangerous. If automated, machines can use several mechanical processes and separation techniques. Examples of mechanical

Fig. 12.1 Landfills contain different types of materials that need to be sorted first

processes are milling, sieving, and crushing; examples of separation techniques are using magnets, air currents, water flow, and optical viewers. Also, artificial intelligence can be used to sort waste. After sorting and separating, glass, metals, paper, and some plastics can be used to create new raw materials. These raw materials can be used to produce new products.

12.1.1 Using Artificial Intelligence to Sort Waste

To be able to automate separating waste from landfills, it is necessary to first identify wasted objects and what materials they are made of so that a robot can later separate the waste into different containers. For object and material identification, artificial intelligence can make separating a lot easier. And better!

Before artificial intelligence can be used to separate waste, it first has to be trained to recognize objects and materials in the **training phase**. Such training involves several steps:

- Step 1: The first step is to take many pictures of all sorts of waste. Most (not all!) images will be used to train the neural network to recognize different object features. Neural networks in artificial intelligence are inspired by the working of a human brain, so they can be thought of as an artificial brain.
- Step 2: These pictures have to be categorized into different types of materials, e.g., paper, plastic, glass, and organic. This must be done by hand so that the neural network can learn this as well.
- Step 3: To make sure an intelligent scanner is later able to recognize different versions of the same piece of waste, data augmentation is used (Fig. 12.2). This means that basic transformations such as rotation are applied to the images. If this step is left out, later recognition would be poor as the neural network would be too strict.

Fig. 12.2 Data augmentation means that basic operations, such as rotation (*middle picture*) or distortion (*right picture*) are applied to improve recognition

Fig. 12.3 A computer doesn't see a banana, it only sees pixels (*right picture*)

- Step 4: Then the neural network looks for basic properties of the images, such as lines, borders, contours, textures, and shadows. These properties are the basic features that the scanner will later be looking for. This may sound easy, but is not trivial: a computer cannot recognize objects as we do, and only sees tiny dots called pixels (see Fig. 12.3). This means the scanner needs to be able to recognize which pixels make up a line, a border, etc.
- Step 5: After that, the complexity of the picture needs to be reduced, so that only the most important information is kept.
- Step 6: This information is then combined in a property list and stored in the neural network.

To speed up this process and improve the training, it is also possible to use existing neural networks. Such networks have already learned from millions of images and therefore have optimized the characteristics as described in step 4. And just like we don't only look at basic properties such as the contour of a face, these networks are also able to process images on a deeper and more abstract level. For example, recognizing different objects within the same image.

After the training is done with most of the images, the remaining images are used for the **validation phase**. The validation must be done using images that the neural network has not seen before, so that the results are due to artificial intelligence, not due to recognition.

With this technique, more than 85% of the materials such as paper, plastic, glass, and organic can be recognized correctly. This is much better than we humans can do, even though some materials cannot be recognized correctly yet. Especially distinguishing between plastic and glass is not always easy. But as neural networks can learn, the stored characteristics are constantly updated and improved. And as artificial intelligence can be combined with mechanical methods, future waste—including glass and plastic bottles—can be separated even better.

12.1.2 Waste Glass as Example of a Valuable Resource

One material that can be extracted from landfills is glass waste. Glass waste is a great resource that can be put to good use in many different ways. As glass in landfills often contains a mixture of colored glass, transparent glass, and other types of waste, it is unsuitable to use for new glass. But glass waste, especially when it is crushed, can still be used for many different purposes.

One example of putting crushed glass waste to good use is by creating **glascrete**. Glascrete is a mixture of concrete and glass. Adding glass particles for example between 75 µm and 2.36 mm (0.003–0.09 in) improves concrete, as the compressive strength becomes up to 43% higher, both initially and in the long-term. Compressive strength is the load the glascrete can carry before it cracks and is the most important characteristic. The optimal proportion of crushed waste glass for an increase in compressive strength depends on glass particle size, texture, how much cement is replaced by glass, and what type of cement is used. As a consequence, it is important to do a trial first before glascrete is produced on a larger scale. In the video in Fig. 12.4 is explained how you can create your own glascrete for pavement stones.

Another example of putting crushed glass waste to good use is by creating **glasphalt**. Similar to glascrete, it is a mixture, in this case of asphalt and glass. For asphalt, the stiffness instead of compressive strength is the most important characteristic. With 15% crushed glass waste, the stiffness is optimal, and also the skid resistance and amount of light reflection, which makes driving safer.

A third example of putting crushed glass waste to good use is by creating **bricks**. In bricks, adding up to 30% crushed waste glass to clay is fine as the bricks meet minimum quality standards or even improve the bricks with higher compressive strength and water absorption. When adding more, the compressive strength of the bricks reduces and they break more quickly. In the video in Fig. 12.5, you can see how glass is used to create bricks in Zanzibar.

A fourth example of putting crushed glass waste to good use is by creating **tiles**. In tiles, it is a great partial replacement for porcelain or feldspar, which is a mineral that forms rocks, but only in small quantities. This is because too much crushed waste glass makes the material for example shrink and more porous. With the right amount of waste glass, the tiles can be improved, because for example porosity is reduced and the material breaks less quickly when it expands due to heat. In the video in Fig. 12.6 you can see how such tiles can be used for the floor in a house.

Fig. 12.4

Fig. 12.5

Fig. 12.6

A fifth example of putting crushed glass waste to good use is by creating **foam glass**. Foam glass is a lightweight and strong material that doesn't chemically react with other materials and doesn't burn. It can be used in many different ways, for example as a lightweight alternative for concrete in buildings and as a thermal insulator. It is one of the most economical ways to put huge amounts of waste glass to good use and can also be used for tiny particles smaller than 0.053 mm (0.002 in). Also, the production process requires a lot less energy compared to cement-based alternatives and the result is environmentally friendly as well. The video in Fig. 12.7 shows how this foam is produced.

A sixth example of putting crushed glass waste to good use is by using it for **water filtration**. The goal of water filtration is to remove solid substances and impurities from contaminated water so that the water can be used for other purposes. Very small particles are used, similar to sand, from fine of 0.2 mm (7.9 μin) to coarse particles up to 2.5 mm (0.04 in). When these particles are used instead of sand, the water filtration works just as well and sometimes even better. Also, 20% less glass waste than sand is needed, which reduces the cost.

A seventh example of putting crushed glass waste to good use is using it for **sandblasting** (see Fig. 12.8). The goal of sandblasting is to make a surface rough or clean by spraying sand under high pressure. Even though using crushed waste glass for sandblasting is still being developed, glass particles of 1.68 mm (0.07 in) or smaller seem to work well with steel. In the video in Fig. 12.9, you can see how glass is used for sandblasting.

Fig. 12.7

Fig. 12.8 This person is sandblasting

12.1.3   Energy Generation

The second way to turn excavated waste into tomorrow's resource is by using materials that cannot be recycled to generate energy. Energy can be recovered using two methods.

The first method to recover energy involves making holes in the waste pile so that methane and other gases can be collected (see Fig. 12.10). These gases are produced by decaying materials. For example, a power plant in Italy is producing 3.8 megawatts of electricity from gas produced by an old landfill. This is similar to the electricity used by an average European house in one year. Once fewer gases are produced, it is safe to start the second method.

The second method to recover energy involves excavating and burning materials in a process called thermovalorization. During thermovalorization, materials that can't be recycled are used as fuel in an electricity-producing plant. For example, 6000 tons of waste per day in a power plant in Bogotá (Columbia) produce 150

Fig. 12.9

Fig. 12.10 Gases from landfills are captured and used to generate electricity

megawatts every day. And Sweden applies this method for materials that can't be recycled anymore so that only 1% of trash ends up in landfill. They even import trash from other countries.

12.2 Conclusion

So, while landfills are growing in size and number, a lot of valuable resources remain unused. Luckily, excavated waste from landfills can become tomorrow's resource by putting it to good use.

As landfills contain a mix of many different materials, it is important to separate waste first if waste is to be recycled. This can be done using mechanical methods such as milling, sieving, and crushing. And it can be combined with artificial intelligence to make separation even more accurate and effective. Once valuable materials have been excavated, they can be used for other purposes. For example, excavated glass waste can be used in concrete, asphalt, bricks, tiles, and foam glass, and be used for water filtering and sandblasting.

Alternatively, landfills can be used to generate electricity. This electricity can be generated by putting the gases produced by landfills to good use or by burning unrecyclable waste. Sweden shows that this can lead to only 1% of waste being sent to landfill.

12.3   How We Can Take Action

As landfills have such a large negative impact on the environment, it is important to limit landfills. Here are practical ideas of what you and I can do to reduce the amount of waste that is sent to landfill:

- Buying in a packaging-free shop
- Separating waste in the correct recycling bin
- Using reusable instead of single-use products
- Bringing waste to recycling
- Using glass instead of sand for sandblasting
- Creating your own garden pavement stones with recycled glass
- Using waste materials for crafting instead of buying new materials, for example using an old glass as salt shaker (see Fig. 12.11)

Fig. 12.11 An old glass used to create a salt shaker

Credit

This Chapter Is Based On:

Landfills as Resource:
Pecorini, I., & Iannelli, R. (2020). Characterization of excavated waste of different ages in view of multiple resource recovery in landfill mining. *Sustainability, 12*(5), 1780.

Artificial Intelligence:
Náñez Alonso, S., Forradellas, R., Morell, O., & Jorge-Vázquez, J. (2021). Digitalization, circular economy and environmental sustainability: The Application of Artificial Intelligence in the Efficient Self-Management of waste. *Sustainability, 13,* 2092.

Glass Waste:
Kazmi, D., Williams, D. J., & Serati, M. (2020). Waste glass in civil engineering applications—A review. *International Journal of Applied Ceramic Technology, 17*(2), 529–554.

Figure Credits

Fig. 12.1   vchal on Shutterstock
Fig. 12.2   Dr. Erlijn van Genuchten
Fig. 12.3   Dr. Erlijn van Genuchten
Fig. 12.8   N_Sakarin on Shutterstock
Fig. 12.10 Roman Mikhailiuk on Shutterstock
Fig. 12.11 Dr. Erlijn van Genuchten

Part III
Biodiversity

While the number of humans on our planet increases, biodiversity decreases dramatically: already 83% of wild animals and half of the plant biomass have gone. Biodiversity is the variety of animals and plants that live in a certain area. This is critical and has caused the biodiversity crisis, as biodiversity is essential for keeping ecosystems healthy. An ecosystem is a biological community consisting of organisms that interact with their physical environment. Biodiversity loss is critical because the animals and plants that are gone were part of natural processes that kept our ecosystems stable; without their presence, natural processes change.

For example, a type of animal that we are losing rapidly is insects; in Germany, the number of insects has reduced by 75% over the last 30 years! And also in other areas, from the Arctic to the tropics. One of the reasons is that we are using pesticides and insecticides in agriculture extensively. But as insects play an important role in fertilizing flowers so that plants can bear fruits, their rapid loss is worrisome.

Fig. 1 Manual pollination is required when there are not enough pollinators

In some places, their numbers have gone down so drastically that pollination has stopped altogether and people pollinate by hand! (Fig. 1)

This shows—and more examples are described in the upcoming chapters—that biodiversity loss has far-reaching consequences and that taking this crisis lightly can be devastating! The good news: there is a lot we can do to make a positive change too.

Credit

Figure Credits

Fig. 1   PattyPhoto on Shutterstock

Chapter 13
How Amazon Deforestation Affects Biodiversity

Abstract The Amazon rainforest, a crucial ecosystem for our planet's ecosystem and biodiversity, is facing significant deforestation, which means that many trees are being cut down. This deforestation has far-reaching consequences as it impacts vegetation and animals, disturbing the natural balance of the ecosystem. Vegetation is affected by changed plant species composition, leading to savannization and degraded ecosystems. Animals, such as the harpy eagle, are also affected by deforestation as they hunt and nest in high trees; when 70% of the trees are cut down, they struggle to survive as they can't find enough food or reproduce. This can even lead to species going extinct.

Keywords Science · Science communication · Biodiversity · Biodiversity loss consequences · Amazon rainforest · Deforestation · Harpy eagles · Savannization · Plant species composition · Agroforestry system · Climate change · Forest degradation

The loss of large parts of the Amazon rainforest in recent decades is becoming a huge concern. This is because this rainforest plays a very important role in our planet's ecosystem. Our planet's ecosystem consists of all living organisms, the environment, and their interactions. For example, the Amazon rainforest is considered the lungs of our planet because it converts a lot of CO_2 from the atmosphere into oxygen. Also, it is very important for biodiversity because of the huge number of species that live and grow in this area.

The Amazon rainforest is located mostly in Brazil (60%), with the other 40% being located in eight other countries: Bolivia, Colombia, Ecuador, Guyana, Peru, Suriname, Venezuela and French Guiana (see Fig. 13.1).

The size of the Amazon rainforest was 526 million hectares in 2020, which is about 9.5 times the size of France! But because of deforestation, its size has been

---

Credit: This chapter is based on two scientific articles by Diego Oliveira Brandão and Everton B. P. Miranda and colleagues. (Full citations are available at the end of the chapter)

Fig. 13.1 The location of the Amazon rainforest

Fig. 13.2 Illegal deforestation of the Amazon rainforest

and is still decreasing, even though the rate reduced substantially since the beginning of 2023. Deforestation means that forest is turned into another land cover, such as agricultural fields or infrastructure including buildings and roads, or that the area is covered with trees by less than 10% for a long time. This differs from cutting single trees or cutting a single tree species from a forest on a large scale, as this is considered forest degradation. Deforestation can both be legal and illegal (see Fig. 13.2).

Fig. 13.3

As deforestation of the Amazon rainforest contributes to economic growth, huge areas are cleared every year. On average, about 500 million trees and palms on 13,856 km² have been cut down every year between 1988 and 2020, which means that in this time frame, a rainforest area of 80% of the size of France got lost. Especially on the eastern and southern edges, which is clearly visible in the timelapse video in Fig. 13.3.

Cutting down so many trees has huge consequences for the first and third planetary crises: climate change and biodiversity loss. This is how deforestation of the Amazon rainforest affects biodiversity in and around the rainforest:

13.1   Vegetation

The first way deforestation of the Amazon rainforest affects biodiversity is by impacting vegetation. The impacted vegetation does not only involve the trees that are cut down, but also vegetation that is left untouched:

13.1.1   Native Species

The first reason vegetation is affected by Amazon rainforest deforestation is that native species have a harder time thriving. Every year, over half a billion native trees and palms are being lost! This decrease in the number of native trees and palms can be temporary when new trees are planted, but persistent if the cleared ground is used for other purposes. For example, the occurrence of the native Brazil nut (see Fig. 13.4) is expected to decrease by 25% by 2050.

The loss of native species has a significant impact on the structure and functioning of the natural Amazon rainforest ecosystem. For example, with fewer native trees being present, also the number of their seeds decreases. With fewer native seeds being dispersed, seed dispersal of exotic species becomes more likely. An exotic species is a species that does not naturally grow or live in a certain area. This causes a further threat to the Amazon rainforest because exotic species affect ecosystems in several ways, including mutual dependencies between plants and animals. A mutual dependency means that plants and animals depend on each other,

Fig. 13.4 A Brazil nut tree (*Bertholletia excelsa*) which is native to the Amazon

such as plants offering fruits to animals so that they have something to eat and animals in return spreading the seeds of these plants to support the plant species' survival (further reading: Chap. 14). This means that native species not only have a harder time to thrive because they are cut down, but also because exotic species are more likely to disturb natural balance.

13.1.2   *Plant Species Composition*

The second reason vegetation is affected by Amazon rainforest deforestation is that it changes plant species composition. Plant species composition means how much each plant species adds to the total vegetation. For example, it is estimated that about 417 million of the roughly 390 billion trees in the Amazon rainforest are Brazil nut trees, which means this species makes up about 0.1% of the tree species composition.

Fig. 13.5 A Pau-de-lacre (*Vismia guianensis*) plant

After deforestation, the plant species composition changes dramatically. This is because the environmental circumstances in deforested and abandoned areas are completely different. As a result, different species—independent of whether they are native or exotic—can thrive. Which species can thrive depends on soil characteristics, including how compact and fertile the soil is. Also, the composition is influenced by whether animals feed on plants' seeds, whether humans manage the area or not by growing new plants, and how well plants can spread.

For example, the number of *Vismia* plants, such as Pau-de-lacre (see Fig. 13.5), increases on degraded lands because it multiplies aggressively from the stems and roots. With its growth, it competes with native or cultivated plants for light, water, nutrients, and space. As a result, the plant species composition changes even further.

13.1.3 Savannization

The third reason vegetation is affected by Amazon rainforest deforestation is that it causes savannization. Savannization means that forest vegetation is slowly being replaced by savannah-like open spaces. These savannah-like open spaces are degraded ecosystems with very few trees.

Savannization is caused by deforestation because cutting down trees changes the Amazon rainforest's climate. These changes are brought about by differences for example in how much sunlight is reflected into space and how much water evaporates from soil and plants into the atmosphere. Also, replacing forests with agricultural land can reduce annual rainfall by 9–25%, extending the length of the dry season. As a consequence, species that require a wet climate die and species that can survive in drier savannah-like conditions survive.

Fig. 13.6   Harpy eagle which has caught a bunny

13.2   Animals

The second way deforestation of the Amazon rainforest affects biodiversity is by impacting animals. How animals are impacted differs between species, as animals benefit from the rainforest in different ways.

One of the animals that is highly impacted by deforestation of the Amazon rainforest is the harpy eagle (see Fig. 13.6). Their lives are intertwined with the rainforest's trees as they hunt and nest in the high trees. When 70% of the trees are cut down, the adults can't survive. This is why:

13.2.1   Hunting

The first reason harpy eagles are affected by deforestation is because their hunting becomes problematic. This is because they are very large birds, weighing up to 7.5 kg (16.5 lbs), and need to eat 800 g (1.8 lbs) of food every day to stay healthy. To get enough food, they prefer sloths and large primates, but also eat large lizards and birds. They find these large prey animals among the trees. The animals in deforested areas are much smaller, so provide much less food per hunt.

Also, the vision of harpy eagles is designed to hunt in the upper tree canopy. They can easily spot their prey from high above the trees, as their prey is large, but have trouble seeing the smaller animals on the grasslands in deforested areas. So, when trees are removed, their prey is not only less nutritious, but it is also harder to hunt. When they find less food less often, they can starve.

Fig. 13.7 Harpy eagle protecting their chick

13.2.2 Nesting

The second reason harpy eagles are affected by deforestation is because their nesting grounds are being removed. Normally, harpy eagles spend their entire adult lives of around 50 years close to their nest tree and the surrounding area. When building a nest, a young couple first locates a suitable tree, a very tall t-shaped tree in the middle of the thick rainforest (see Fig. 13.7). They use this nest for the rest of their lives, raising one chick every three years. When trees are cut down, the adults need to find a new tree, but when they can't find one, they are unable to raise new chicks.

13.3   Conclusion

So, while trees in the Amazon rainforest are being cut down at an alarmingly fast rate, the consequences on biodiversity are clearly noticeable and far-reaching. These consequences are noticeable for both vegetation and animals.

Vegetation is impacted by deforestation because native species have a harder time surviving, the type of plants that grow in this area changes, and the area is subject to savannization. These changes have far-reaching consequences for the rainforest's ecosystem as land degrades, and fewer and different plants can survive.

Also, animals are impacted by deforestation because for example their hunting and nesting grounds change or become unavailable. This in turn can cause animals to starve or have difficulties reproducing. This eventually can cause species to go extinct. If animals reduce in numbers, this also impacts predators that feed on them, causing the whole food chain to shift.

13.4　How We Can Take Action

As the Amazon rainforest including its biodiversity is so important for our planet, it is important to protect whenever we can. Here are practical ideas of what you and I can do to protect the Amazon rainforest:

- Reducing paper and wood use
- Reducing meat consumption as part of the agricultural land is used to grow crops to feed cattle
- Supporting initiatives that contribute to saving the Amazon rainforest
- Supporting indigenous communities that play an important role in conserving the Amazon rainforest
- Voting for politicians that support its conservation

Credit

This Chapter Is Based On:

Vegetation:
Brandão, D. O., Barata, L. E. S., & Nobre, C. A. (2022). The effects of environmental changes on plant species and forest dependent communities in the Amazon region. *Forests, 13*(3), 466.

Harpy Eagles:
Miranda, E. B. P., Peres, C. A., Carvalho-Rocha, V., Miguel, B. V., Lormand, N., Huizinga, N., Munn, C. A., Semedo, T. B. F., Ferreira, T. V., Pinho, J. B., Piacentini, V. Q., Marini, M., & Downs, C. T. (2021). Tropical deforestation induces thresholds of reproductive viability and habitat suitability in Earth's largest eagles. *Scientific Reports, 11*(1), 1–17.

Figure Credits

Fig. 13.1　Shubhamtiwari on Shutterstock
Fig. 13.2　PARALAXIS on Shutterstock
Fig. 13.4　"'Bertholetia excelsa' Castanha-do-Pará Brazil-nuts tree" by mauroguanandi is licensed under CC BY 2.0 DEED
　　　　　Source: https://commons.wikimedia.org/wiki/File:%22Bertholetia_excelsa%22_Castanha-do-Par%C3%A1_Brazil-nuts_tree.jpg
　　　　　Author: https://www.flickr.com/photos/mauroguanandi/4657870321/
　　　　　License: https://creativecommons.org/licenses/by/2.0/deed.en
Fig. 13.5　"Vismia guianensis (Aubl.) Choisy" by Alex Popovkin, Bahia, Brazil is licensed under CC BY 2.0 DEED
　　　　　Source: https://commons.wikimedia.org/wiki/File:Vismia_guianensis_(Aubl.)_Choisy_(15488221924).jpg, Author: https://www.flickr.com/people/12589168@N00
　　　　　License: https://creativecommons.org/licenses/by/2.0/deed.en
Fig. 13.6　Chepe Nicoli on Shutterstock
Fig. 13.7　feathercollector on Shutterstock

Chapter 14
How Exotic Animal Species Harm Ecosystems

Abstract Globalization is causing interactions between different parts of the world, which occasionally involves the—accidental or intentional—introduction of animal species in other regions. These so-called exotic species can significantly impact biodiversity by changing the structure and functioning of native natural ecosystems. Exotic animal species, such as rabbits, parrots, and long-tailed macaques, affect native plant species by changing seed dispersal and mutual dependencies between plants and animals. This can lead to reduced seed diversity and reduced survival chances for plant species. Also, exotic animal species can affect seed quality by harming seeds or removing essential nutrients, affecting their ability to grow into plants.

Keywords Science · Science communication · Biodiversity · Biodiversity loss consequences · Seed dispersal · Ecosystems · Invasive species · Ecosystem functioning · Exotic species · Biological invasions · Dispersal disruption · Plant-animal · Mutualism · Seed predation

While native species are at risk in the Amazon rainforest due to deforestation because exotic species join the ecosystem, this trend also happens in other parts of the world. This is caused by globalization. Globalization means that we start operating on an international scale. This trend not only causes our influence to reach out to other parts of the world but also the influence of other species. This can happen when we, for example, (accidentally) import seeds or animals when we fly to a different part of the world.

When species start living in other parts of the world, they are called exotic species in these areas. As exotic species, they can be a huge threat to biodiversity. This is because they change the structure and functioning of natural ecosystems in these

---

Credit: This chapter is based on the scientific article "Ecological Impacts of Exotic Species on Native Seed Dispersal Systems: A Systematic Review" by Sebastián Cordero, Francisca Gálvez, and Francisco E. Fontúrbel. (Full citation is available at the end of the chapter)

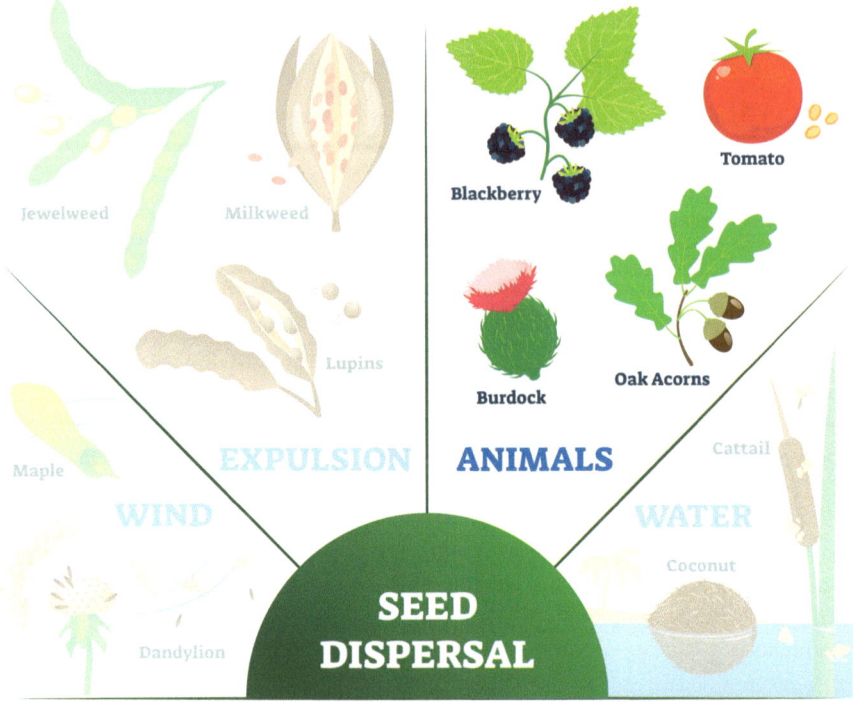

Fig. 14.1 Seeds can be dispersed by different forces. Here, we focus on animals

regions. When they cause harm to the natural ecosystem, they are called invasive species.

For example, exotic animal species change how seeds are spread. These changes can have a big impact as 90% of plants in tropical regions depend on animals to spread their seeds and 60% in non-tropical regions (see Fig. 14.1). This is how exotic animal species affect ecosystems by changing seed dispersal:

14.1   Affecting Native Plant Species

The first way exotic animal species impact ecosystems through changed seed dispersal is by affecting native plant species. One example of an exotic animal species that affects native plant species is the European rabbit. This rabbit is a good seed spreader and supports seeds' growth. But unfortunately, it is also good at spreading seeds of invasive plants and eats native tree seedlings.

Another example of an exotic animal species that affects native plant species is the parrot. They spread both native and exotic plants when seeds stick to them or when eating fruits and pooping out the seeds.

Fig. 14.2 Long-tail macaques

A third example of an exotic animal species that affects native plant species is the long-tailed macaque (see Fig. 14.2). It spreads but also destroys native seeds. This is because they more often discard unripe native fruits than unripe exotic fruits. When ripe fruits are available, they prefer exotic fruits.

14.2   Affecting Mutual Dependencies

The second way exotic animal species impact ecosystems is by affecting mutual dependencies between plants and animals. A mutual dependency for example exists when plants offer fruits to animals so that they have something to eat, and in return, animals spread the seeds of the plant by carrying them away. Both providing food and spreading seeds are important for these species' survival (see Fig. 14.3).

This mutual dependency is affected when exotic species are introduced into an area. One reason this dependency is affected is that exotic plants compete with native plants for animals to eat their fruits. As a consequence, animals visit native plants less often.

Another reason is that the diversity of seeds being spread reduces. This is for example possible when exotic animals hunt and eat native animals. As a consequence, these native animals have to move to a different habitat or die. In both cases, they can't spread seeds in their initial habitat anymore.

Apart from these direct consequences, the loss of these mutual dependencies has also many indirect consequences. For example, changed mutual dependencies impact whether plant species can survive or go extinct. In case they go extinct, they can't provide other services anymore either, such as shelter and shadow.

Fig. 14.3 Animals and plants live in a mutual dependency, which means they depend on each other for survival

14.3 Affecting Seed Quality

The third way exotic animal species impact ecosystems is by affecting seed quality. One reason seed quality is affected by exotic animal species is that while some spread seeds, others harm seeds by consuming or manipulating them. When seeds are harmed, they may not be able to grow into a plant.

Another reason seed quality is affected is that exotic animal species sometimes remove parts of the seed. For example, some species remove the parts of seeds that attract ants (see Fig. 14.4). These ants would normally carry the seeds to their nest as these parts contain important nutrients. Without these parts, ants won't spread these seeds.

14.4 Conclusion

So, exotic animal species impact ecosystems by affecting native plant species, mutual dependencies between animals and plants, and seed quality. In many cases, this impact is negative. And while in some cases the impact seems to be small, it is also important to consider indirect effects. This is because indirect effects can be far-reaching and could be negative in the long run. For example, when a few plants are spread disproportionately wide, diversity reduces. With fewer different species, a disease affecting a habitat can do more harm than with more different types of plants.

Fig. 14.4 Seeds with extruding parts that are attractive to ants

14.5   How We Can Take Action

As introducing exotic species can have far-reaching, negative consequences, it is important to limit their introduction in other parts of the world. Here are practical ideas of what you and I can do to prevent the negative impact of exotic animal species:

- Refraining from bringing animals when traveling
- Refraining from bringing exotic animals back home as pet
- Planting native species in garden
- Checking whether plants in garden are native species and if not, removing them
- Refraining from feeding exotic animals
- Controlling exotic species if they can't be removed, for example by governments providing hunting permits

Credit

This Chapter Is Based On:

Cordero, S., Gálvez, F., & Fontúrbel, F. E. (2023). Ecological impacts of exotic species on native seed dispersal systems: A systematic review. *Plants*, *12*(2), 261.

Figure Credits

Fig. 14.1   VectorMine on Shutterstock
Fig. 14.2   rujithai on Shutterstock
Fig. 14.3   RLS Photo on Shutterstock
Fig. 14.4   Korean botanist on Shutterstock

Chapter 15
How Climate Change Affects Ocean Biodiversity

Abstract Human activity, such as cutting trees and introducing animals to other parts of the world, affects biodiversity. Not only directly but also indirectly. Indirect effects are possible because everything is connected in ecosystems, and changes in one part of an ecosystem can cause changes in other parts. For example, our actions cause climate change which in turn heavily affects ocean biodiversity. This is because climate change causes our oceans to warm up, become more acidic, and contain less oxygen. This in turn impacts marine species in multiple ways, both on an individual and population level, including species' survival and reproduction.

Keywords Science · Science communication · Biodiversity · Biodiversity loss consequences · Climate change · Ocean biodiversity · Ocean warming · Ocean acidification · Ocean deoxygenation · Planetary health · Natural systems · Human systems

The previous two chapters clearly show how human activity, such as cutting trees and introducing animals to other parts of the world, can have far-reaching consequences on biodiversity. Unfortunately, our actions also indirectly affect biodiversity. This is because everything is connected in ecosystems and every part—even the smallest creature—plays an important role. As a consequence, changes in one part of the ecosystem cause changes in other parts of the ecosystem. This in turn means that our planet is not suffering from three *distinct* environmental crises (climate change, pollution, and biodiversity loss) but three *interrelated* environmental crises and that these crises can fuel each other.

For example, climate change affects biodiversity. One ecosystem in which biodiversity is heavily affected by climate change is our planet's oceans. Oceans make up the core of our planet's hydrosphere, which consists of all waters on the earth's

---

Credit: This chapter is based on the scientific article "Climate change-accelerated ocean biodiversity loss & associated planetary health impacts" by Byomkesh Talukder, Nilanjana Ganguli, Richard Matthew, Gary W. vanLoon, Keith W. Hipel, and James Orbinski. (Full citation is available at the end of the chapter).

surface. Because of this major role, biodiversity changes and losses in these habitats can have major impacts on the world. And if we continue emitting carbon like we do today, our oceans could even begin to release carbon back into the atmosphere, which will make both the biodiversity and climate change crises worse! This is how climate change affects ocean biodiversity:

15.1 Ocean Warming

The first way climate change affects ocean biodiversity is through ocean warming. Ocean warming means a rising ocean temperature. In Fig. 15.1, you can see that the temperature of our oceans is increasing as the temperature in 2023 (bold orange line) is much higher than the average temperature between 1982 and 2011 (bold black line). The thin gray lines show variations in the temperature in different years during the last decade. Note that they are all above the average of 1982–2011.

This temperature rise is caused by greenhouse gases, which prevent heat from being radiated back into space. As a consequence, this heat remains in the atmosphere; 90% of this heat is in turn absorbed by our oceans. This additional heat affects organisms both on an individual and population level.

On an individual level, warmer temperatures affect marine species in multiple ways. These include:

- their digestion and immune system can be impacted
- they can get breathing problems
- the leaf width of aquatic plants can be affected
- plants can become more likely to have epiphytes, which are plants that grow on other plants
- larvae are less likely to survive

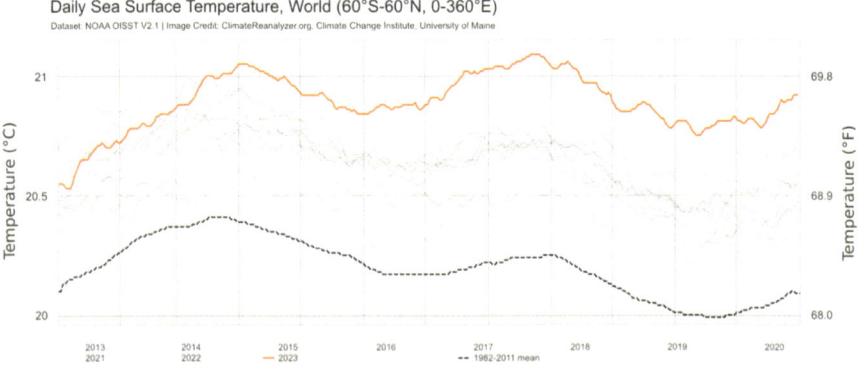

Fig. 15.1 Graph showing the ocean surface temperature around the world between 2013 and 2023 The thin *gray lines* indicate the temperature in single years; the *bold dotted line* indicates the average sea/ocean temperature between 1982 and 2011; the *orange line* indicates the temperature in 2023

Fig. 15.2 A coral reef that is bleaching due to warmer water.

- plants can deal with environmental stresses less well
- coral reefs bleach (see Fig. 15.2)

On a population level, warmer temperatures affect marine species in multiple ways as well. These include:

- the number of different species can change
- the size of communities can reduce even leading to species going extinct
- certain species migrate towards the poles. Whether species decide to migrate depends on factors such as the temperature they can cope with, how their habitat changes due to higher temperatures, how much food is available, and whether invasive species have entered their original habitat.

15.2   Acidification

The second way climate change affects ocean biodiversity is through acidification. Acidification means that our oceans become more acidic. This happens mainly because oceans absorb additional CO_2 due to large amounts of CO_2 being emitted into the air by burning fossil fuels. As CO_2 chemically reacts with water molecules this results in a higher concentration of hydrogen ions (H^+).

This higher concentration of hydrogen ions has caused the acidity (pH) to increase since the beginning of the Industrial Revolution over 200 years ago. The pH scale goes from 0 to 14, where 7 is neutral. Values below 7 are considered acidic and values above 7 are basic or alkaline (see Fig. 15.3). The current pH level of our

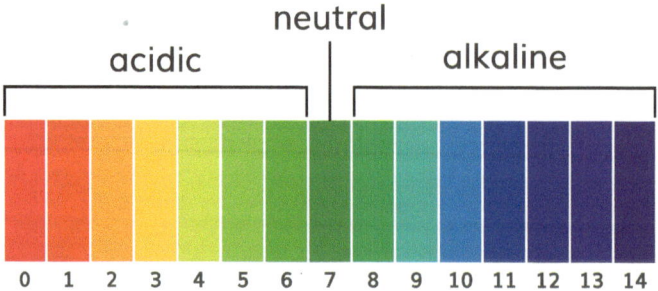

Fig. 15.3 The pH scale

oceans is 8.1, which means it is alkaline. At the beginning of the Industrial Revolution, this was 8.2, so the pH level has decreased by 0.1 since then. This does not sound much but as the pH scale is not linear but logarithmic, it means that the acidity has increased by about 30% during that time. This higher acidity affects organisms both on an individual and population level.

On an individual level, higher acidity affects marine species in multiple ways. These include:

- reproduction can be affected, for example because sperm doesn't move as well and fast
- the development and growth of organisms can be affected
- the metabolism of organisms can be affected. Metabolism refers to the chemical reactions in cells to provide energy and grow new cells
- breathing rates can change
- organisms can die younger
- shells can disintegrate

On a population level, acidification affects marine species also in multiple ways. These include:

- it can make calcareous species less competitive because the anatomic structure is impacted. Calcareous species are species containing calcium carbonate such as snails with shells. This changes ecosystems and eventually results in lower biodiversity
- the number of individuals in a population can reduce

The animation in Fig. 15.4 nicely explains acidification and its far-reaching consequences.

15.3 Deoxygenation

The third way climate change affects ocean biodiversity is through deoxygenation. Deoxygenation means that water can retain less oxygen. This happens because of several reasons:

Fig. 15.4

Fig. 15.5

- the warmer water caused by climate change can store less oxygen
- warmer water stays close to the surface and doesn't mix as well with water further down from the surface. This means that also oxygen can't be transported to deeper levels as well
- the exchange of oxygen between the ocean and the atmosphere depends on organisms living in the ocean that perform photosynthesis. These organisms are threatened by ocean warming so that less exchange takes place
- organisms living in the ocean need more oxygen in warmer water because their metabolisms are higher

On an individual level, deoxygenation affects marine species by making embryos less likely to survive. On a population level, deoxygenation affects marine species by changing species interactions. For example, being exposed to low oxygen levels for a short while impacts predator-prey interactions.

The video in Fig. 15.5 also nicely explains deoxygenation and its consequences.

15.4   Conclusion

So, human activities can have indirect consequences on biodiversity, as our actions cause anthropogenic climate change which in turn affects biodiversity. In our oceans, climate change affects biodiversity by causing the water temperature to increase, the water to become more acidic, and the amount of oxygen in the water to reduce. This affects both individual marine plants and animals, and whole populations. It for instance reduces their survival chances and reproduction. As all parts of ecosystems are connected and play an important role, this in turn causes further changes, such as changes in the food chain and food availability.

Fig. 15.6

15.5   How We Can Take Action

As environmental crises are interrelated and ocean biodiversity is affected by CO_2 emissions, it is helpful to limit harmful activities as much as possible. Here are practical ideas of what you and I can do to prevent further ocean warming, acidification, and deoxygenation:

- Urging the government to take action on climate change and ocean conservation, for example by voting for leaders who take climate change seriously and take action
- Reducing your energy consumption at home to decrease your carbon footprint, for example by insulating the house and only heating as much water as needed
- Eating less meat and opting for meat alternatives instead, such as burgers made from soybeans or even printed meat
- Supporting organizations that protect our oceans and marine species, such as restoring coral reefs. In the video in Fig. 15.6, you can virtually visit one of these sites.

Credit

This Chapter Is Based On:

Talukder, B., Ganguli, N., Matthew, R., vanLoon, G. W., Hipel, K. W., & Orbinski, J. (2022). Climate change-accelerated ocean biodiversity loss & associated planetary health impacts. *The Journal of Climate Change and Health*, *100*, 114.

Figure Credits

Fig. 15.1   "Daily Sea Surface Temperature, World (60°S-60°N, 0-360°E)" by Climate Reanalyzer is licensed under CC BY 4.0 DEED / Fahrenheit scale added
            Author: https://climatereanalyzer.org/
            Source: https://climatereanalyzer.org/clim/sst_daily/#info
            License: https://creativecommons.org/licenses/by/4.0/
Fig. 15.2   Ethan Daniels on Shutterstock
Fig. 15.3   AlexVector on Shutterstock

Chapter 16
How Whales Change Our World

Abstract The planet is facing three interrelated environmental crises, with climate change affecting biodiversity and biodiversity loss contributing to climate change. Whales are important for biodiversity as they control marine animal populations and move around nutrients. And even after they have died, they are a source of nutrients for decades. Also, they contribute to mitigating climate change; directly because they store carbon for centuries and indirectly because their existence allows kelp forests to thrive, which capture a lot of carbon as well. As a consequence, restoring whale populations—which were significantly reduced over the last 1000 years—can move the biodiversity and climate change crises into a positive spiral.

Keywords Science · Science communication · Biodiversity · Biodiversity loss consequences · Climate change · Marine ecosystem · Cetaceans · Great whales · Migration · Habitat · Baleen whale · Sperm whale

The previous chapter clearly showed that our planet is not suffering from three *distinct* environmental crises but three *interrelated* environmental crises as these crises can fuel each other. While climate change can affect biodiversity, biodiversity loss can also contribute to climate change. This is for example because oceans have removed 25% of all carbon emissions in the last 60 years from the atmosphere through natural processes. But these natural processes are threatened by biodiversity loss.

Some of these natural processes are heavily influenced by whales. Whales (*Cetaceans*) live life on a scale that we can hardly imagine: they live in unknown ocean depths, travel nearly the length of the world from feeding grounds to birthing grounds each year, and some are one of the largest animals ever to live on Earth (see Fig. 16.1). The hugest whales are collectively called "great whales". Great whales include the Sperm Whale and 16 species of Baleen Whales. Baleen Whales don't

---

Credit: This chapter is based on two scientific articles by Joe Roman and Celine van Weelden and their colleagues. (Full citations are available at the end of the chapter).

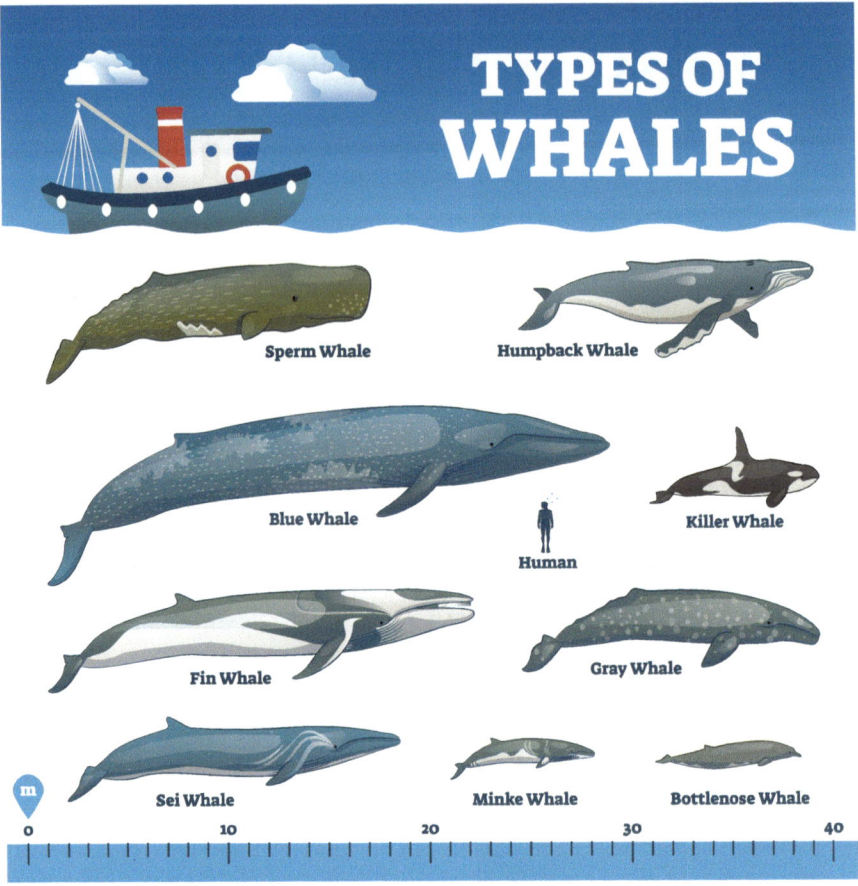

Fig. 16.1 Different types of whales and their length in meters

have teeth and filter plankton from the water. Their size ranges from 6 m (about 20 ft) for the Pygmy Right Whale to 31 m (about 102 ft) for the Blue Whale. And the Blue Whale weighs nearly 200,000 kg (441,000 lbs)! That is similar to 2500 average-weight men from North America!

In the past, these giants were common. But this changed about 1000 years ago when people in Spain started hunting Right Whales. Whales were sold for their meat to eat, for their oil to be used during cooking and for other purposes, and for baleen to be used in clothing as it is strong and flexible. Baleen is the part in their mouths that filters plankton (see Fig. 16.2). Over the next 1000 years, whale populations were reduced by as much as 90%, causing great whale species to become endangered, and some even close to being extinct.

Although hunting has largely been stopped, climate change now affects whale populations. For example:

• some whale species have changed their living habits because of weather and water temperature shifts. For example, Humpback, Blue, Fin, and Sperm whales

Fig. 16.2 The brown "hairs" in this whale's mouth is baleen

have all moved towards the poles in search of cooler waters. Their presence reduces the habitat available for Bowhead, Narwhal, and Beluga Whales.
• others have changed the timing of their migrations, including the Humpback, Fin, and Blue whales, who now spend more time in high latitudes.

As these trends continue, some of these species, like the Bowhead, may benefit and increase in numbers, while others, like the Beluga and Narwhals, may become extinct due to loss of habitat.

Obviously, these trends are life-changing to the great whales themselves. What is less obvious, is that these trends are also indirectly life-changing for us. This is because the whales play important roles in the stabilization of oceans, and in turn, the stabilization of the world's climates. And as climate change and biodiversity are influencing each other, biodiversity changes and losses can end up in a downward spiral, causing further and wide-ranging consequences for us and our planet. This is why whales are important to biodiversity and climate change – to the future of our planet's oceans and us:

16.1   Whales as Consumers

The first reason whales are important to our planet's oceans is because they consume a lot of food. By eating so much, they control and stabilize the populations of certain marine animals. These animals, often small in size such as krill and plankton, quickly increase or decrease in numbers in response to changes in their

Fig. 16.3

environment. For example, changes in our planet's climate, number of predators, and available food supplies can lead to large changes in their populations. This is possible because small animals usually have short lives, and short lives often go together with high fertility and a lot of offspring.

Great whales can control those populations by eating excess populations and moving on when the populations are small. In the video in Fig. 16.3 is explained that eating small amounts is not worth the effort. This is because when eating small populations, they need more energy to speed up again after eating than they consumed: opening their mouth slows them down a lot and because they are so heavy, they need a lot of energy to accelerate.

With great whales decreasing in numbers, their ability to stabilize and control marine animal populations is reduced directly. Also, it has indirectly affected other marine animal populations and how much CO_2 is captured. For example, killer whales—who hunt great whales—reduced in numbers or had to change their diet to other species such as harbor seals, sea lions, and sea otters. This in turn reduced the populations of these smaller animals, which allowed a population boom in sea urchins. Sea urchins eat kelp (see Fig. 16.4), a type of seaweed. The increase in sea urchin populations depleted kelp forests and reduced the fish populations living in these forests. As kelp can capture CO_2 and store carbon for a long time, smaller and fewer kelp forests contribute to climate change.

16.2   Whales Move Around Nutrients

The second reason whales are important to our planet's oceans is because they move around nutrients. Not only do they move nutrients from the bottom to the surface by eating deep down and pooping at the surface, they also move nutrients in the water around by diving. This is because their diving moves deep water full of nutrients upward. For example, during hunting, Humpback whales plow the bottom of the ocean looking for prey. This releases stored minerals and nutrients. Also, Humpback whales create fish nets made from bubbles. These bubbles not only capture their prey fish, but they also move nutrient-rich water to the surface. Such a bubble net is shown in the video in Fig. 16.5.

Apart from moving nutrients from the bottom to the surface, whales move nutrients from the poles to the tropics. Pregnant female whales store nutrients while

Fig. 16.4 A kelp forest

Fig. 16.5

being in the feeding zones near the poles. When they migrate to the birthing grounds, they normally fast, using the nutrients they have stored to feed themselves and their baby. In this case, they don't let go of poop, but urine instead, also including iron and nitrogen.

16.3 Whales as Nutrient Sources

The third reason whales are important to our planet's oceans is because they are important nutrient sources. One important source is their feces, their poop: it is filled with nutrients that are hard to find in oceans, especially near the surface. Whales often expel their feces near the ocean's surface. The most important minerals that they expel are nitrogen and iron. These nutrients are essential for many ocean inhabitants near the surface, including plankton.

Fig. 16.6

Another important source is their bodies. During their lives, they are prey to other animals. This may sound surprising because usually, animals as large as great whales have no natural predators. But killer whales, or orcas, do hunt large whales. This can be recognized by scrapes and marks on the flukes of most large whale species.

Also, after they have died and their bodies have sunk to the ocean floor, their bodies are a very important nutrient source for other animals during different decay phases:

1. First, sharks, hagfish, and other fish eat the soft tissues of the carcass.
2. After that, the nutrients that have ended up in sediments and the fat-rich bones are eaten by a wide range of small creatures and bacteria.
3. In the last stage, chemical scavengers, such as sulfides, dissolve the remaining skeleton.

This process can take many decades and continuously supplies nutrients during these stages. These phases are explained and shown in the video in Fig. 16.6.

16.4   Whales as Carbon Storage

The fourth reason whales are important to our planet's oceans is because they capture and store carbon. They consume carbon through their food and store it in their bodies as blubber. Blubber is their body fat. Carbon stored in blubber is kept out of the atmosphere for the rest of the whale's life. This means this carbon is locked away for centuries. Depending on the species, this can be up to 100 years.

But also after whales die, the carbon remains captured: whale carcasses carry about 190,000 tons of carbon a year from the atmosphere to deep waters. As whales are good at storing carbon, they are great at mitigating climate change. Each great whale stores 33 t (about 73,000 lbs) of CO_2 on average per year. To compare, a tree stores only up to 22 kg (48 lbs) of CO_2 a year. That is over 1500 times as much!

16.5   Conclusion

So, whales are very important to our world because they are critical to our planet's oceans and marine animals. This is because they directly impact this ecosystem by controlling marine animal populations and moving around nutrients. And even after they have died, they are a source of nutrients for decades and store carbon for centuries.

Also, they indirectly impact this ecosystem and our world's climate. This is because their existence has far-reaching consequences for other parts of the marine ecosystems through the food chain. For example, fewer whales eventually leads to fewer kelp forests being able to store carbon.

16.6   How We Can Take Action

As whales play such an important role in our world and are essential for biodiversity and mitigating climate change, it is critical to support them. Here are practical ideas of what you and I can do to help whales thrive in the deep oceans:

- Supporting international regulations to stop all hunting of great whales
- Supporting whale research organizations, through time or money
- Talking to your friends about whale poop
- Refraining from hunting whales
- Refraining from buying products that are made from whales, such as whale meat
- Reducing our carbon footprint to reduce the rising temperatures in oceans by replacing fossil fuels by renewable energy sources such as wind and solar energy
- Disposing of plastic properly and reducing plastic use to prevent whales from eating it

Credit

This Chapter Is Based On:

Roman, J., Estes, J. A., Morissette, L., Smith, C., Costa, D., McCarthy, J., ... & Smetacek, V. (2014). Whales as marine ecosystem engineers. *Frontiers in Ecology and the Environment, 12*(7), 377–385.
van Weelden, C., Towers, J. R., & Bosker, T. (2021). Impacts of climate change on cetacean distribution, habitat and migration. *Climate Change Ecology, 1*, 100,009.

Figure Credits

Fig. 16.1   VectorMine on Shutterstock
Fig. 16.2   John Tunney on Shutterstock
Fig. 16.4   Madelein Wolfaardt on Shutterstock

Chapter 17
Biodiversity Solutions: Protecting Natural Ecosystems

Abstract As biodiversity is crucial for our survival and even contributes to mitigating the climate change crisis, it is critical to protect biodiversity. One way to preserve biodiversity is by protecting and restoring natural ecosystems. We can protect and restore natural ecosystems by reducing the amount of land that we convert into urban or farm areas, repairing critical habitats, removing pollutants from the environment, and managing forests. This involves for example converting biodiversity-poor areas into biodiversity-rich areas, repairing critical habitats by reintroducing species, restoring wetlands, banning harmful products, preventing environmental pollution, limiting threats that harm forests, and maximizing the benefits of forests. Together, these contribute to thriving natural ecosystems.

Keywords Science · Science communication · Biodiversity · Biodiversity loss solutions · Natural ecosystem restoration · Forest management · Plant pathogens · Forestry · Habitat reduction · Habitat restoration · Pollutant removal · Tree disease · Epidemiology · Pathogen · Invasive species · Species diversity

As biodiversity is critical for our world and our survival, and biodiversity loss fuels the climate change crisis, it is important to take action urgently. One way to take action is by protecting natural ecosystems. This is helpful because healthy ecosystems allow different plant and animal species to thrive. This includes humans: even though many feel that nature is something outside of us, we are part of nature and ecosystems! And we depend on natural processes, just like other living creatures. For example, we depend on natural processes that provide us with oxygen to breathe, clean water to drink, and food to eat.

Even though many of our activities—such as cutting trees in the Amazon rainforest and urbanization—are harmful to natural ecosystems, luckily, it is still possible

---

Credit: This chapter is based on two scientific articles by Phoebe Barnard and Michaela Roberts and their colleagues. (Full citations are available at the end of the chapter)

to reverse these negative effects by protecting and restoring them. This is how we can save our planet by protecting and restoring natural ecosystems:

17.1   Stopping Habitat Reduction

The first way to protect and restore natural ecosystems is by stopping the conversion of wild areas into urban or farm areas. Wild areas are, for example, forests, grasslands, coastal mangroves, marshes, and ocean kelp forests. Protecting these areas is an essential part of increasing CO_2 removal and protecting biodiversity. This can be done by for example:

- **Protecting Green Areas** Green areas within cities need protection as they provide a habitat for animals and remove air pollution. But they also provide other benefits, including shade and cooling, food, and space for leisure activities.
- **Reducing Urban Expansion** New city policies and bylaws can reduce urban expansion by encouraging denser populations in the city center. Although regional and national policies are important, cities can often make faster progress on issues that affect the local area.
- **Converting Biodiversity-Poor Into Biodiversity-Rich Areas** Some green areas, such as lawns and parking spaces, hardly contribute to biodiversity and carbon absorption. Converting these underused spaces into healthy biodiverse spaces will contribute to restoring natural ecosystems (see Fig. 17.1).

Fig. 17.1 Leaving part of a lawn untouched contributes to restoring ecosystems

17.2   Repairing Critical Habitats

The second way to protect and restore natural ecosystems is by repairing critical habitats. Critical habitats are areas that are important for biodiversity. They for example contain endangered species and species that are essential for a healthy food chain. This can be done by for example:

- **Reintroducing Species** By reintroducing predators and other species that stabilize food chains in critical habitats, the balance in the ecosystem can be restored. This is because top predators, such as lions and great whales, ensure a balance between their prey and other species.
- **Restoring Wetlands** By restoring wetlands (see Fig. 17.2), critical wildlife habitat is restored, as wetlands are for example important breeding grounds and sources of food for many animals, including fish, birds, and amphibians. They also control erosion and flooding, protect and improve water quality, and capture and store carbon.

Fig. 17.2 Wetlands contain important ecosystems

17.3 Removing Pollutants

The third way to protect and restore natural ecosystems is by removing existing pollutants from habitats, especially wetlands, soil, and air. As pollution often stays around for a long time, it continues to do damage; removing those pollutants and preventing new pollutants from being added will stop additional damage. This can be done by for example:

- **Banning Polluting Products** Products that are being used but harm the environment should be prohibited. For example, lead shot has for many years been used to hunt animals. But as lead is highly toxic, it kills many animals yearly. And even we are in danger, as lead can reach us through the food chain and harm us too, such as causing Parkinson's disease (further reading: Chapter 10 of A Guide to a Healthier Planet Volume 1: "How Heavy Metal Pollution Can Cause Parkinson Disease")
- **Removing Polluting Products** For example plastic often ends up in the environment, where it stays for a long time and harms ecosystems (further reading: Chapter 8 of A Guide to a Healthier Planet Volume 1: "How Plastic Pollution Impacts Aquatic Animals"). To stop the harm done, it is important to prevent plastic from ending up in the environment in the first place, and remove it when it is already there (see also Chapter 11 "Pollution Solutions: Removing Plastic Waste From The Environment").
- **Removing Air Pollution** As air pollution is hard to remove, it is best to prevent pollutants from reaching the atmosphere in the first place. The remaining air pollution can be removed with for example filtration systems, plants (see Fig. 17.3),

Fig. 17.3 Houses with green walls reduce air pollution

and using nano-additives on buildings (further reading: Chapter 12 of A Guide to a Healthier Planet Volume 1: "Pollution Solutions: Removing Pollutants From Air").

As the cost of removing pollution is becoming more clear, it is also becoming clear what the true costs are of the products we buy. This understanding hopefully helps reduce unnecessary consumption and waste.

17.4   Managing Forests

The fourth way to protect and restore natural ecosystems is by managing forests. Forests are maybe more important to us than we realize. This is because they benefit the ecosystem and other organisms in many different ways. Forests are for example called the lungs of our planet because they clean the air we breathe. They also filter the water we drink, prevent erosion, and help capture and store CO_2, which helps reduce global warming. And they are large homes and natural refuges for many different types of plants and animals. Also, forests contain many natural resources, such as timber, that help local communities to thrive, and medicinally important plants.

These benefits exist, independent of whether forests grow naturally or are grown by humans, for example for commercial purposes. But they don't only have the same benefits, they also share risks. Whereas only natural forests are at risk due to for example cutting trees, urban development, and poaching, both natural and man-grown forests are affected by natural threats that negatively impact the trees' health.

Managing forests involves limiting threats and supporting benefits:

17.4.1   Limiting Threats

Limiting threats means detecting, preventing, and responding to threats. This can be done by for example:

- **Increasing Plant Biodiversity** Increasing plant diversity in forests is important as it protects a forest against diseases. This is because microorganisms that harm only particular species have fewer plants to grow on, the distance between plants of the same species is larger so that microorganisms are less likely to spread to the next plant, and more organisms that naturally control the pest find a place to live in the forest.
- **Reducing Wood Burning** Reducing wood burning is especially important when wood is used as a replacement for fossil fuels, as burning wood produces more CO_2 than burning coal. It also causes a lot of air pollution. And as a tree is cut down, the capacity of the forest to absorb CO_2 is reduced because fewer trees can absorb CO_2. Also, other plants are disturbed when trees are cut down.

- **Limiting Tree Harvesting** Limiting the number of trees that are harvested as much as possible is important, as simply replanting the trees doesn't compensate for the loss of CO_2 absorption. This is because undisturbed forests contain plants, animals, insects, fungi, lichens, and soil full of bacteria and viruses that double the amount of CO_2 that can be absorbed. Also, replacing harvested forests with seedlings reduces CO_2 absorption, as young trees only absorb a fraction of the CO_2 absorbed by a mature tree.
- **Controlling Diseases** One of the natural threats is plant diseases caused by microorganisms. Microorganisms include bacteria, fungi, viruses, and microscopic worms (see Fig. 17.4). Most of these harmful organisms are invasive species causing diseases. But as these microorganisms are so small and their presence is often hidden from view until they cause major damage (see Fig. 17.5), controlling them is a challenge. This is similar to cancer cells: in the beginning, we don't realize they are there until they have already done major damage. Also, as different microorganisms can cause the same symptoms, recognizing the cause is not always easy.
- **Distinguishing Between Primary and Secondary Infection** Forest management practices are also important to find out whether the infection is primary or secondary. Primary infection is when the plant is affected by a harmful microorganism for the first time. Secondary infection is when the plant has been affected before. This is important to know, as primary infections show that a new harmful microorganism has entered the forest. As harmful microorganisms are often noticed late and different microorganisms cause the same symptoms, it is important to identify early on which species causes the harm. This is possible by looking at the reproductive structures of the organisms, such as spores. Also, for even more reliability, the genetic makeup can be identified with their DNA.

Fig. 17.4 A microscopic worm

Fig. 17.5 Microorganisms can cause harm to trees

- **Preventing Disease Spreading** When trees or other plants are affected by diseases, it is important to prevent diseases from spreading. For example, when a tree is cut down because of illness, the remaining tree stump and roots should be removed as well so that no harmful microorganisms and their spores are left behind. This can be done using mechanical methods such as raking, or chemical methods. Plant leftovers can also be burned, but this also causes damage to other plants and animals. The positive effects of stump removal can be seen up to 21–50 years after taking action!

17.4.2 Maximizing Benefits

While managing forests involves limiting threats, it also involves maximizing the benefits. This can be done by for example:

- **Site Preparation** To keep forests healthy, correct site preparation is important. Site preparation involves all earthwork done to prepare and maintain the forest. For example, site preparation can involve using fertilizers. In some cases, this is helpful to prevent tree illness, as more tree nutrients increase the trees' strength and reduce the impact of harmful organisms.
- **Increasing Plant Diversity** Increasing plant diversity in a mixed culture is also important to keep forests healthy. Plant diversity is higher in a mixed culture than in a monoculture, because a mixed culture contains two or more plant species, and a monoculture only one (see Fig. 17.6). Increasing the diversity can happen naturally, but can also be done on purpose by planting different species. When

Fig. 17.6 A mixed culture makes a forest more likely to stay healthy

done on purpose, it is important to make sure the selected plants are not suitable alternatives for the harmful microorganisms, so that microorganisms have fewer plants to grow on. Also, the distance between plants of the same species is larger, so that microorganisms are less likely to spread to the next plant.

- **Providing Plant Shelter** Providing plant shelter to the next generation of trees that thrive in areas with a lot of shade also benefits forests. This is especially helpful to protect young trees from harmful microorganisms or spores that travel through the air, carried by the wind. While being protected by larger, stronger plants, they are less likely to get airborne diseases. Also, growing in a protected environment, the physical strength and health of small trees improve, so that they are more resistant to harmful microorganisms. Trees that thrive in areas with a lot of sunlight probably do better without plant shelter.
- **Controlling Tree Density** Another method to keep forests healthy is controlling tree density. Tree density means how closely together trees grow: with high density, trees grow closer together than with low density. Controlling tree density means making sure that trees have a suitable distance from each other. This is important to prevent harmful microorganisms from spreading between trees. Especially when these microorganisms can spread through the roots. To prevent spreading, trees can be removed or fewer trees can be grown from the start.

- **Thinning** Thinning is an important practice to keep forests healthy. Thinning means removing plants from the forest. Plants chosen for thinning are dead or dying plants, plants with a high risk of becoming infected, and plants with disease symptoms. One advantage of thinning is that by removing these plants, both plants of the same species and plants of other species are protected from becoming ill. This is because harmful microorganisms are less likely to spread. Another advantage of thinning is that the reduced forest density makes it possible for other plants to grow stronger. When they are stronger, they are more resilient against diseases and more likely to stay healthy. This also means that the quality of the wood is higher when trees are used for timber.
- **Pruning** Apart from thinning a forest, it is helpful to prune plants (see Fig. 17.7). Pruning means removing parts of plants. One advantage of pruning is that the humidity in the forest can be reduced. Lower humidity can prevent harmful microorganisms from thriving. Another advantage is that infected parts of the plants can be cut away before more damage is done.
- **Limiting Connectivity** Another method to keep forests healthy is limiting connectivity. Connectivity involves to what extent forests or forest patches are connected. A connection exists for example when roots are connected, spores blow over, water carries diseases, humans or animals carry harmful microorganisms, and vehicles and tools carry infected soil. The closer a healthy and diseased forest are together, the more likely it is that the healthy forest gets infected too. So, keeping an appropriate spatial distance between forest patches and using sanitary precautions for the tools and vehicles are important to limit connectivity.

Fig. 17.7 Pruning trees can be helpful to keep the forest healthy

17.5   Conclusion

So, to support biodiversity and limit the negative impact of the biodiversity crisis, it is helpful to protect and restore natural ecosystems. We can do this for example by stopping habitat reduction, repairing critical habitats, removing pollutants from the environment, and managing forests.

While all these activities are important, managing forests is probably more important than we realize. This is because forests are not just contributing to biodiversity, they also help mitigate the climate crisis. Managing forests involves limiting threats and maximizing benefits. Limiting threats encompasses for example increasing plant biodiversity, reducing wood burning, limiting tree harvesting, controlling diseases, distinguishing between primary and secondary infection, and preventing disease from spreading. Maximizing the benefits encompasses for example site preparation, increasing plant diversity, providing plant shelter, controlling tree density, thinning, pruning, and limiting connectivity. Together, they contribute to a thriving forest ecosystem.

17.6   How We Can Take Action

As protecting and restoring natural ecosystems is important for biodiversity, it is important to make a difference in daily life whenever possible. Here are practical ideas of what you and I can do to protect natural ecosystems:

- Growing plants on balconies in urban areas
- Looking for a place to live within an urban area instead of converting a natural area to build a house
- Growing plants in your garden instead of maintaining a lawn; alternatively reducing the size of the lawn by partly replacing grass with other plants
- Growing different types of native instead of non-native species in your garden
- Refraining from using polluting products, such as non-biodegradable plastics
- Disposing of polluting products such as non-biodegradable plastics properly, so that they don't end up in nature
- Preventing air pollution, for example by cycling or using public transport instead of driving a car to work
- Refraining from burning wood as fuel, especially when it is used as a replacement for fossil fuels
- Participating in local cleanups of critical areas, such as wetlands
- Removing soil from shoes before entering a forest

Credit

This Chapter Is Based On:

Ecosystems:
Barnard, P. et al. (2021). World scientists' warnings into action, local to global. *Science Progress*, *104*(4), 00368504211056290.

Forests:
Roberts, M., Gilligan, C. A., Kleczkowski, A., Hanley, N., Whalley, A. E., & Healey, J. R. (2020). The effect of forest management options on forest resilience to pathogens. *Frontiers in Forests and Global Change*, *3*, 7.

Figure Credits

Fig. 17.1   Dr. Erlijn van Genuchten
Fig. 17.2   Irina Wilhauk on Shutterstock
Fig. 17.3   Sommart sombutwanitkul on Shutterstock
Fig. 17.4   F.Neidl on Shutterstock
Fig. 17.5   Tricky_Shark at Shutterstock
Fig. 17.6   Panga Media on Shutterstock
Fig. 17.7   Thomas Soellner on Shutterstock

Chapter 18
Biodiversity Solutions: Sustainable Fishing

Abstract Food security is at risk due to the growing global population and while eating fish is a solution to address hunger, it can also cause biodiversity issues when practiced unsustainably. To ensure that we can continue to enjoy the benefits of the fishing industry, this industry needs to be made more sustainable. One way to achieve this is by minimizing the unintentional catch of marine animals, for example, by implementing sensory cues as deterrents. Another way to achieve this is by using artificial intelligence to address challenges such as weather changes, navigation issues, pirate attacks, and technical issues. A third way to achieve this is transforming fish waste into a valuable resource by putting it to good use instead of throwing it away.

Keywords Science · Science communication · Biodiversity · Biodiversity loss solutions · Fishing industry · Sustainable fishing · Fishing · Artificial intelligence · Fish waste · Bycatch · Bycatch mitigation · Sensory deterrents · Sea turtle · Elasmobranch · Seabird · Marine mammal · Boat automation · Surveillance · Machine learning · Fish byproduct valorization · Marine sustainable sources · Collagen · Bioactive peptides · Chitin · Oil · Enzymes

With the growing global population and the wide-ranging consequences of the three planetary crises, food security is becoming an increasing issue—even though the United Nations set the goal to end hunger in the world by 2030. To resolve world hunger, eating fish is proposed as a solution because eating fish and even fish waste provides essential nutrients, including vitamins, minerals, and omega-3 fatty acids.

But while eating fish contributes to the solution of food security, fishery can also cause many biodiversity issues when practiced in an unsustainable way. One of the reasons is that fish nets are often left behind in the ocean, causing animals to get caught and get harmed or die (see Fig. 18.1). Such ghost nets make up about 10%

Credit: This chapter is based on three scientific articles by Sol Lucas, Rajakannu Amuthakkannan, and Daniela Coppola and their colleagues. (Full citations are available at the end of the chapter)

Fig. 18.1 Many marine animals get trapped in ghost nets

of all marine litter. It is estimated that each year between 500,000 and one million tons of ghost nets and other fishing equipment are added.

Another reason is that fishery puts ecosystems at risk. This is because marine animals that are caught can't contribute to the marine food web anymore. The marine food web consists of all food chains in the marine ecosystem. And unfortunately, fishers don't only catch fish for us to eat, they also find a lot of bycatch in their nets. Bycatch means the unwanted fish and other marine animals that get trapped in (commercial) fishing nets and traps while trying to catch a different species. Most animals become bycatch because they get entangled or are attracted via sensory cues. For example, animals can see or go after the smell of prey in and around gear, and maybe even try to eat them.

So, to preserve biodiversity, sustainable fishing is essential. Sustainable fishing means that enough fish are left in the ocean, marine habitats are respected, and the livelihoods of people who work in the fish industry are ensured. This is how the fishing industry can be made more sustainable:

18.1   Preventing Bycatch

The first way to make the fishing industry more sustainable is by preventing bycatch. The consequences of catching fish and having bycatch are widespread. It for example causes food chains to collapse and food webs to change, it threatens both commercial and small-scale fisheries and communities that depend on this fishery, and it can put species at risk. For example, fishing for knife fish has been prohibited in the Chinese Yangtze River, so that these fish can recover.

As the consequences are large, it is very important that as few animals as possible die unnecessarily. This means that the amount of bycatch needs to be reduced. To achieve this, many different solutions have been proposed and put in place.

One solution has been prohibiting highly destructive ways of fishing that cause a lot of bycatch. For example, dynamite fishing, also called blast fishing or bomb fishing, is not allowed anymore. It involves using dynamite or other explosives to stun or kill fish (see Fig. 18.2). As they sink to the bottom or float to the surface after being harmed, they can be caught easily. But also when using fishing nets and other gear, bycatch is common. How many animals end up as bycatch is not exactly clear, as a lot is illegal, unregulated, and unreported catch.

Other solutions focused on changing fishing gear, releasing bycatch after capturing them, reducing fishing, applying catch limits, and prohibiting fishing in specific areas for a certain time. But these solutions are often unpopular with fishers as they not only reduce bycatch but can also reduce the target catch. Or animals are harmed, even though they can escape. That is why it is important for fish not to get into touch with gear in the first place.

To make sure fish don't get in touch with gear in the first place, several solutions have been developed that use sensory cues. These are sensory cues that can be used to deter bycatch:

Fig. 18.2 Coral reef with dead fish after destructive dynamite fishing

18.1.1 Acoustic Deterrents

The first type of sensory cues that can prevent bycatch are acoustic cues. Acoustic means through hearing. Acoustic deterrents can be used to trigger adverse responses in *sharks*:

- Artificial sounds between 20 Hz and 20,000 Hz can be sent to prevent sharks from catching prey. These frequencies are in the range of what humans can typically hear, with lower frequencies being lower tones. This is helpful but acoustic cues need to be combined with cues for other senses to be successful.
- Apart from using artificial sounds, sharks can also be deterred using replay sounds. For example, a wild killer whale's call can be used to deter white sharks. But this works well only when a call was recorded in for example Australia and also replayed in Australia. It didn't work when being replayed in for example South Africa. This could be because killer whales' calls vary between regions and whales are only triggered by the variation they know.

Acoustic deterrents can also be used to trigger adverse responses in *sea birds*. Pingers that emit a sound of 1500 Hz can for example be used to reduce the bycatch of common murres by half but it didn't work with rhinoceros auklets (see Fig. 18.3).

Acoustic deterrents can also be used to trigger adverse responses in *marine mammals*:

- Alarms with frequencies between 40,000 and 160,000 Hz and pingers with frequencies ranging from 10,000 to 160,000 Hz can reduce the bycatch of harbor porpoises (see Fig. 18.4) when these are attached to gillnets. Gillnets are fishing

Fig. 18.3 Murres (*Uria aalge*) can be deterred using pingers to prevent them from being killed unnecessarily (*left*) whereas rhinoceros auklet birds (*Cerorhinca monocerata*) cannot (*right*)

Fig. 18.4 A porpoise (*Phocoena phocoena*)

nets that are hung vertically in the water so that fish get trapped in them by their gills (see Fig. 18.5).

- Pingers with 10,000 Hz were also helpful in strongly reducing dolphin bycatch: about 86% less! These pingers should be used with care though because they can exclude animals from their habitat. Or they can be like 'dinner bells' for some species, meaning that they are attracted by the pinger, causing more bycatch. For example, while the bycatch of common dolphins and northern elephant seals (*Mirounga angustirostris*) reduced, the bycatch of California sea lions increased. That is why it is helpful to have different options and decide for different situations which solution is the most effective.

18.1.2   Olfactory Deterrents

The second type of sensory cues that can prevent bycatch are olfactory cues. Olfactory means through smelling. Olfactory deterrents can be used to trigger adverse responses in *sharks*:

- Semiochemicals from dead sharks, also called necromones, can be used to repel sharks. Semiochemicals are chemicals that are released by animals in the environment to send a signal to another organism to change its behavior. When smelling these chemicals, they decided to temporarily stay away.
- The type of bait can also reduce shark bycatch, for example using squid instead of fish. But the effectiveness depends on species-specific food preferences.

Olfactory deterrents can also be used to trigger adverse responses in *seabirds*:

Fig. 18.5 A gillnet

- Decomposing animal flesh leftovers can be used to deter certain birds, such as albatrosses, from eating bait from or settling on longlines.
- Shark liver oil can be used to prevent flesh-footed shearwaters (see Fig. 18.6) and other seabird species from diving behind vessels.

How successful using these smells is, depends on the bird species.

Olfactory deterrents can also be used to trigger adverse responses in *turtles*. Replacing squid as bait with fish bait decreases the number of turtle bycatch. For example, the number of loggerhead (see Fig. 18.7) and leatherback (see Fig. 18.8) turtles that were accidentally caught were a lot smaller when using mackerel as bait.

18.1.3   Visual Deterrents

The third type of sensory cues that can prevent bycatch are visual cues. Visual means through seeing. Visual deterrents can be used to trigger adverse responses in *sharks*:

- Green LEDs seem to reduce shark bycatch a lot, but this has to be studied in more detail as the results come from relatively few comparisons.
- The SharkSafe barrier can be used to deter sharks, as this barrier is clearly visible underwater. In the video in see Fig. 18.9, you can see what it looks like and how a shark swims away.

Fig. 18.6 A flesh-footed shearwater (*Ardenna carneipes*)

Fig. 18.7 A loggerhead turtle (*Caretta caretta*)

Visual deterrents can also be used to trigger adverse responses in *seabirds*:

- Turning off the light on vessels that use longlines for fishing seems to reduce the amount of seabird bycatch. This is because fewer birds are active in the dark or cannot find hooks with bait only based on smell.
- Using scaring lines, or tori lines (see Fig. 18.10), also reduces the number of seabird bycatch, although other factors such as weather conditions, line quality,

Fig. 18.8 A leatherback turtle (*Dermochelys coriacea*)

Fig. 18.9

and the height of these lines also influence their effectiveness. These lines are above the water.

- When using drift gillnets, which are gillnets that are kept floating at the proper depth, it is helpful to make the upper 20–50 meshes white. These white meshes are a visual alert and reduce bycatch of for example common murres by 40–45%.
- Coloring bait blue can also help reduce bycatch, but the disadvantage is that it can also reduce the target catch.
- Whether using laser beams is effective is unclear, because the quality of the studies that found positive effects is questionable. Also, the consequences for birds' eyes need to be studied before these visual cues should be used.
- Orange gillnets seem to deter more little penguins (*Eudyptula minor*) compared to green and clear lines.
- Looming eye buoys are a promising development to deter seabirds. Looming eye buoys are rotating panels with drawn eyes, attached to a buoy (see Fig. 18.11). For example, they significantly reduce the number of long-tailed ducks within 50 m (164 ft) of the buoys.

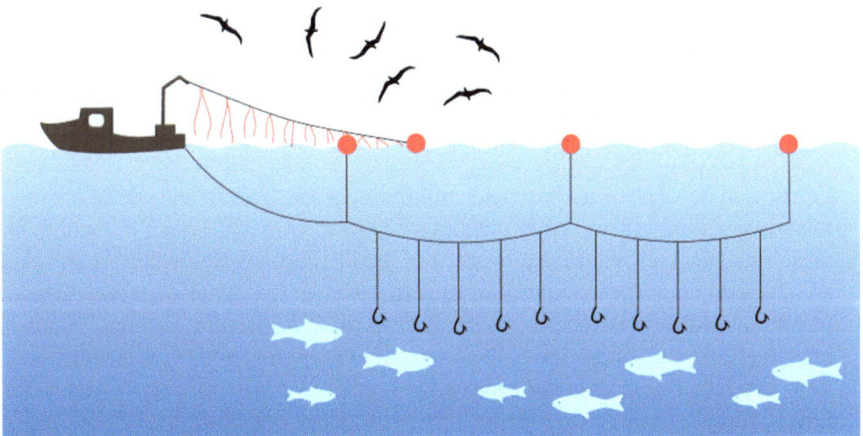

Fig. 18.10 Scaring or tori lines (*red lines above the water*) are used to keep seabirds away

Fig. 18.11 A looming-eye buoy

Visual deterrents can also be used to trigger adverse responses in *marine mammals*. Green LEDs reduced the bycatch of small cetaceans between 67% in nets set on the bottom and 71% in driftnets. Cetaceans are whales, dolphins, and porpoises.

Visual deterrents can also be used to trigger adverse responses in *turtles*:

- LEDs on gillnets that emit ultraviolet light can be used to reduce the bycatch of green turtles by at least 40%, without reducing the amount of target catch.
- 3D shark and 3D sphere models can discourage loggerhead turtles from going after bait.

18.1.4   Electrosensory Deterrents

The fourth type of sensory cues that can prevent bycatch are electrosensory cues. Electrosensory means sensing electrical impulses. Electrosensory deterrents can be used to trigger adverse responses in *sharks*:

- Ferrite magnets can be used to repel multiple species. A ferrite magnet or ceramic magnet is a type of permanent magnet made of mostly iron oxide. For example, these magnets can reduce the bycatch of small shark species when using traps and even increase the amount of target catch.
- Neodymium-based magnets can be used as well to prevent different elasmobranch species, which includes sharks, from looking for food in the fishing location, although the success depends on the species.
- SMART hooks can be used to reduce the bycatch of spiny dogfish (see Fig. 18.12) although bycatch was still high. SMART stands for Selective Magnetic and Repellent-Treated Hook. These hooks combine two repellent methods in standard fishing hooks: magnetism and shark-repellent metals.
- Pulsing electrical and magnetic signals can be used to deter sandbar sharks (*Carcharhinus plumbeus*) and largetooth sawfish (*Pristis pristis*). As these results were obtained in the lab, it still needs to be seen whether these results can also be achieved in real-life fishery.

Fig. 18.12 Spiny dogfish (*Squalus acanthias*)

18.1.5 Echolocation Reflection Deterrents

The fifth type of sensory cues that can prevent bycatch are echolocation reflection cues. Echolocation reflection means reflecting echo signals that are sent by for example dolphins to locate objects. Echolocation reflection deterrents can be used to trigger adverse responses in dolphins. Bottlenose dolphins and harbor porpoises can detect fishing nets from a longer distance. But as nets can become stiffer when materials are added to allow this reflection, results can also be caused by these changes. Also, dolphins living in captivity may react differently than dolphins living in the wild, so more research needs to be done on the effectiveness of echolocation reflection.

18.2 Using Artificial Intelligence

The second way to make the fishing industry more sustainable is by using artificial intelligence. Artificial intelligence is helpful because fishermen are facing a broad range of challenges. These challenges include weather changes, navigation issues, pirate attacks, and technical issues with their boats. As these challenges can cause all sorts of problems such as environmental pollution, and safety and communication issues, addressing them is critical for sustainable development of the fishing industry.

Luckily, the rapid development of modern technologies supports fishermen in addressing these challenges effectively. For example, modern technologies such as robotics and artificial intelligence are already successfully applied in other industries, including the automobile, flight, and manufacturing industries, and can be applied in the fishing industry as well (see Fig. 18.13). By applying these and further advanced technologies effectively, boat automation, information sharing, and documentation can be advanced so that safety, security, navigation, and information sharing can be improved. This is how artificial intelligence can support making the fishing industry more sustainable:

18.2.1 Banning Prohibited Fishing Practices

The first way artificial intelligence can make the fishing industry more sustainable is by contributing to banning prohibited fishing practices by detecting illegal, unreported, and unregulated fishing.

Banning *illegal fishing* involves banning prohibited types of fishing such as bottom brawling and dynamite fishing (see Fig. 18.2). With bottom brawling, big and heavy fish nets are dragged on the sea floor, which catches every marine animal. Both practices cause huge amounts of bycatch and destroy marine habitats. Artificial

Fig. 18.13 Artificial intelligence can for example be used to analyze different types of data from different sources to inform and warn fishermen

intelligence can for example be used to find fishing-related patterns in large data sets so that different types of fishing can automatically be recognized.

Banning *unreported fishing* involves banning fishing activities that are not reported to authorities. These activities are problematic as they limit the effectiveness of national and regional efforts to make fishing sustainable and conserve marine environments. Artificial intelligence can for example be used to ban unreported fishing by identifying and tracking illegal and unknown vessels by analyzing satellite pictures. Often filters are applied to these pictures to for example reduce background noise. This maximizes the difference between the ship and the sea so that ships can be detected more easily. Also, texture differences of the surface can be detected.

Banning *unregulated fishing* involves banning fishing that is not in line with national and regional regulations, for example by fishing during non-fishing months. Fishing during non-fishing months is a threat to marine resources as this prevents marine environments and marine animal populations from recovering. This in turn affects the future of the fishing industry. Artificial intelligence can for example be used to identify and track ships during non-fishing months and automatically warn coastal authorities.

18.2.2   Marime Surveillance

The second way artificial intelligence can make the fishing industry more sustainable is by contributing to maritime surveillance. Maritime surveillance involves monitoring different activities, such as trespassing humans and boats crossing marine borders (see Fig. 18.14).

Fig. 18.14 Marine surveillance can be done using different technologies such as cameras, GPS, and in this picture radar

Surveillance for trespassing humans can for example involve a system used by the government at the port. The system uses cameras, sensors, and artificial intelligence to warn people on the boat and authorities about intruders, and other risks. Also, surveillance for trespassing humans can involve a system that detects and analyzes movements using artificial intelligence. This system should preferably have night vision as well and be able to distinguish between humans and for example smoke. If a human or smoke is detected, the user is warned of the danger.

Surveillance for boats crossing marine borders can for example involve a system with detailed, live GPS tracking. When a boat approaches or crosses a predefined border, a warning is sent to the fishermen and coastline authorities. As crossing the border can be unintentional, this can prevent fishermen from being arrested or even shot. Because of this danger, some systems are even able to shut down the boat's motor when a boundary is crossed. Artificial intelligence can for example be used to prevent border-crossing by reconstructing and forecasting boats' movements, estimating maritime routes, identifying vessel types, and detecting abnormal vessel behavior.

18.2.3   Boat Detection

The third way artificial intelligence can make the fishing industry more sustainable is by contributing to boat detection. Boat detection is—just like surveillance for crossing marine borders—important to avoid conflicts between neighboring

Fig. 18.15 Fishing boats can be equipped with all kinds of technologies to detect other boats

countries. Other boats can be detected using for example the Internet of Things and 360-degree monitoring (see Fig. 18.15). The Internet of Things involves interconnected devices that can send and receive data.

Boat detection using the Internet of Things involves obtaining data from different sensors including GPS, water detector sensors, and ultrasonic sensors. This data is provided to fishermen on a mobile app so that they can prevent collisions, especially with smaller boats that lack relevant technologies. Artificial intelligence can be used to analyze the data from these sources and warn a vessel when unknown objects are identified.

Boat detection using 360-degree monitoring involves cameras to keep an eye on the area around a vessel. This is helpful as humans are limited in doing reliable continuous monitoring, for example, because of physical limitations or misjudgments about the identified vessels. Artificial intelligence can for example be used to analyze data from several cameras that continuously monitor the area around the vessel and measure the distance between identified objects.

18.2.4 Environmental Monitoring

The fourth way artificial intelligence can make the fishing industry more sustainable is by contributing to environmental monitoring. Environmental monitoring can be done using sensor networks and autonomous ships.

Environmental monitoring using sensor networks involves real-time measuring of physical parameters and communicating the gathered data with a central location. Environmental parameters in marine environments are for example water temperature, acidity (pH), and salinity, and oxygen and nutrient density. Such networks are widely used on land while using them for marine environments is still in its infancy. Artificial intelligence can for example be used to predict environmental circumstances in the near future so that necessary preparations or arrangements can be made accordingly.

Environmental monitoring using autonomous ships involves unmanned surface ships, autonomous underwater vehicles, and underwater gliders. Artificial intelligence is used for example for automatic control, to avoid collisions, and to identify targets. Maybe in the future, such devices can—with the use of artificial intelligence—even fish autonomously.

18.2.5   Oil Spill Detection

The fifth way artificial intelligence can make the fishing industry more sustainable is by contributing to oil spill detection. Oil spill detection is important because oil spills can damage fishing engines and are one of the worst environmental disasters. This is because it takes years to clean up oil spills and for marine life to recover. These spills for example come from fishing vessels, cargo ships, and cruise liners (see Fig. 18.16). The earlier these spills are detected, the smaller the environmental damage and the negative impact on fishermen, the fishing industry, and the environment.

Oil spills can be detected using one technology or a combination of technologies. Oil spill detection using a single technology can be done for example by radar technology on satellites or infrared cameras. The advantage of using radar technology is that it can be done 24/7 and also works on rainy and cloudy days. Oil spill detection using a combination of technologies can for example use space, ground-based, and underwater technologies. Artificial intelligence can for example be used to identify oil spills based on data from different sensors and technologies.

18.2.6   Pirate Attack Prevention

The sixth way artificial intelligence can make the fishing industry more sustainable is by contributing to preventing pirate attacks. Pirate attacks on fishing and trade vessels can cause significant harm to the crew of these boats and economic losses. For example, in 2019, 162 boats were attacked by pirates using guns. As artificial intelligence can be used to gain new insights such as the best traveling route, and can communicate insights quickly with authorities on land or with the control rooms of ships, it can also be used for fully automated piracy alerts. Also, data from cameras can be analyzed to predict pirate weapons and alerts be sent accordingly.

Fig. 18.16 Oil spills can harm the environment, marine animals, and fishing vessels

18.3 Putting Fish Waste to Good Use

The third way to make the fishing industry more sustainable is by putting fish waste to good use. This is helpful because the growing world population requires more food to be produced. This increased production means that we need more natural resources: from 1970 to 2017, the global use of natural resources increased by 254%! Unfortunately, the amount of waste increases due to this trend as well.

As this trend of resource use and waste production puts a lot of stress on our planet, it is important to use available resources as efficiently as possible. To achieve higher efficiency, moving towards a circular economy is extremely helpful. A circular economy is an economy in which products are used as long as possible by sharing, reusing, repairing, and refurbishing them, and at the end of their life use their materials for new products. This means that hardly any waste is produced.

An example of a natural resource that has been used more and more in the last few decades is fish: the production has grown more than eightfold between 1954 and 2014! And with it, also the amount of fish waste has increased, as a lot of fish is lost or discarded somewhere in the fishing or production chain. Also, parts of fish are thrown away that we can't eat, such as fish heads (see Fig. 18.17). This increase is so dramatic that it has become an environmental concern. For example, when fish waste is dumped into the ocean, it harms the environment in several ways, including reducing the amount of oxygen in the water, burying or smothering living organisms, and introducing diseases on the sea floor.

Fig. 18.17 Sardines heads as leftovers on a fish market

To be able to reduce the environmental impact, the European Union put a policy in place to significantly decrease the amount of discarded fish waste. At the same time, the goal of this policy is to increase the amount of fish biomass that is used in the best possible way. This best possible way currently means that fish waste is used to create fish oil and fishmeal, which is ground-dried fish used as animal feed or fertilizer. But fish waste can contribute to a circular economy in many more ways, as it contains many nutrients and valuable byproducts. This is how fish waste can be turned into a valuable resource:

18.3.1   Collagen

The first way in which fish waste can be turned into a valuable resource is by extracting collagen. Collagen is a protein that is mainly found in skin and tissues that connect, support, bind, or separate other tissues or organs (see Fig. 18.18). It makes up 20–30% of proteins in animals and when it is treated in a particular way, it becomes gelatin.

Currently, the main sources of collagen are cow and pig skin, cattle bones, and other waste from mammals. But it can also be extracted from bones, skin, scales, or fins of various marine animals including fish, sponges, and jellyfish. Compared to the collagen extracted from cattle, fish collagen is absorbed more efficiently in our body and brought into circulation, so that it can have a larger beneficial effect.

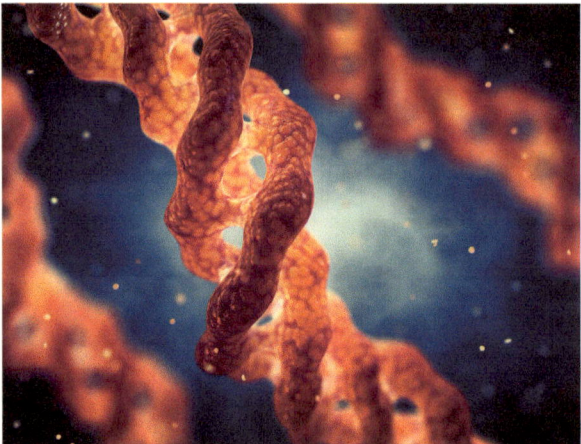

Fig. 18.18 Collagen molecules

Because of its beneficial effect, collagen is often used in the food, beverage, and cosmetic industries. Example applications are:

- tissue engineering: collagen can be used to regenerate cartilage. Cartilage is the strong, flexible connective tissue that protects our bones and joints.
- wound healing: collagen can be used to heal burn wounds, specifically in wet wound dressing, as it promotes several growth factors necessary for tissue repair and prevents infections in the wound
- antioxidative: collagen can protect cells from the damage caused by free radicals. Free radicals are molecules that easily react with other molecules

18.3.2   Bioactive Peptides

The second way in which fish waste can be turned into a valuable resource is by extracting bioactive peptides. Bioactive peptides are small protein molecules that act, for example, as hormones. They can be extracted from bodies or body parts, including frame, scale, bone, head, gonads, and viscera of for example pollack, sole, halibut, ray, tuna, and catfish.

To be able to extract bioactive peptides, the fish waste needs to be treated using different chemical processes. One of these processes is fermentation. Fermentation is the natural breakdown of a substance by bacteria, yeasts, or other microorganisms. Often, enzymatic hydrolysis is used though in food and pharmaceutical industries, because it doesn't leave solvents or toxins behind. Enzymatic hydrolysis is the chemical breakdown of a substance due to its reaction with water with the help of enzymes that speed up this process. After the peptides are extracted, they need to be purified before they can be used.

The extracted and purified bioactive peptides can be used in several ways, including:

- medicines: they can be used in different pharmaceuticals because they for example prevent high blood pressure
- functional food: they can be used in functional foods specialized to help with aging memory deficits because of their ability to protect brain cells
- cosmetics: collagen peptides can be used to support the loss of natural collagen from the skin, improve bone health, and protect against degenerating joints

18.3.3   Fish Oil

The third way in which fish waste can be turned into a valuable resource is by extracting fish oil. This oil can be extracted from almost all parts of the fish, including its flesh, head, frame, fin, tail, skin, and guts. This can be done using different methods, for example wet pressing and chemical extraction. The resulting oil contains omega-3 fatty acids, which are crucial to human health but cannot be produced by our bodies.

The extracted fatty acids and oils can be used for many purposes. It can for example be used as:

- food for aquatic animals: it can be used in aquaculture, which involves rearing aquatic animals or aquatic plants for food
- food supplement: as a food supplement for humans (see Fig. 18.19), it can prevent cognitive decline, reduce blood pressure, prevent developing diabetes, prevent heart and lung diseases, and improve survival rates in cancer patients
- biofuel: it can be used to create biodiesel which can replace fossil fuels

18.3.4   Chitin

The fourth way in which fish waste can be turned into a valuable resource is by extracting chitin. Chitin is a type of sugar without taste or color. It can be extracted from the scales of carp, tilapia, red snapper, and parrotfish. As chitin is naturally surrounded by other molecules, different techniques are used to remove these other molecules. For example, deproteinization is used to remove proteins and can involve washing away the proteins with certain chemicals. From the extracted chitin, a similar molecule called chitosan can be derived.

Chitin and chitosan can be used for many different applications. Chitin can for example be used as a substitute for plastic. Chitosan can for example be used to clean water. And as they kill or slow down microorganisms' spreading, they can be used in open wound treatment and tissue engineering.

Fig. 18.19 Fish oil is full of healthy omega-3 fatty acids

18.3.5   Enzymes

The fifth way in which fish waste can be turned into a valuable resource is by extracting enzymes. Enzymes are substances in a body that speed up chemical reactions, without being changed by these reactions. Some fish' enzymes are unique, as they live in diverse and usually hostile environments. Most of the enzymes can be found in their stomachs, pancreas, and intestines.

The most occurring enzymes in fish are proteases. Proteases are enzymes that break down proteins and peptides. In fish, they are mostly found in the stomachs of codfish, tuna, sardines, salmons, sharks, mackerels, trouts, and other species.

Apart from proteases, fish also contain lipases. Lipases are enzymes that break down fats. They can for example be extracted from Atlantic cod and Atlantic salmon, red sea bream, and rainbow trout (see Fig. 18.3).

Enzymes found in fish are special because they are highly active even in very small concentrations, and they work with low temperatures and a wide acidity (pH) range. That is why they can be used for many purposes. For example:

- in the food and beverage industry to improve flavors, textures, and colors
- in molecular biology research
- in pharmaceutical or agrochemical industries as environmentally safe chemicals for instance used in detergents

18.4 Conclusion

So, as the growing world population puts our food security at risk, it is important to make food production more sustainable. One food industry that can be made more sustainable is the fishing industry. Making this industry more sustainable is helpful because fishing can put a lot of strain on the environment.

Making this industry more sustainable can be achieved in several ways. For example, techniques to deter animals can be used to reduce the amount of bycatch. These techniques trigger acoustic, olfactory, visual, electrosensory, and echolocation senses so that unwanted marine animals stay away.

Also, artificial intelligence can be used to keep an eye on several aspects of fishing practices, such as the presence of oil spills and other boats, and surveillance. This not only benefits the environmental aspects but also the financial and social aspects of sustainability.

Furthermore, fish waste can be turned into a valuable resource instead of throwing it away. It can for example be used as a source for collagen, bioactive peptides, fish oil, chitin, and enzymes. These can benefit our health but also be put to good use in many other contexts.

18.5 How We Can Take Action

As the fishing industry reduces world hunger but making the industry more sustainable is essential, it is important to do as much as possible in daily life to support the industry's sustainable development. Here are practical ideas of what you and I can do to make fishing more sustainable:

- Throwing bycatch back immediately
- Throwing back fish when fishing as a hobby
- Using fishing gear that is suited to target specific species and avoid catching other species
- Avoiding fishing in over-fished areas
- Avoiding fishing in areas where the chance to catch bycatch is large
- Disposing of fishing gear properly instead of discarding it in waterways or the sea

Here are practical ideas of what you and I can do when buying products related to the fishing industry:

- Buying fish from fishermen who apply sustainable fishing practices
- Supporting initiatives that aim at banning prohibited fishing activities
- Buying seafood from local fishermen
- Buying food supplements made from seafood leftovers instead of other sources

Credit

This Chapter Is Based On:

Preventing Bycatch:
Lucas, S., & Berggren, P. (2022). A systematic review of sensory deterrents for bycatch mitigation of marine megafauna. *Reviews in Fish Biology and Fisheries*, 1–33.

Using Artificial Intelligence:
Amuthakkannan, R., Vijayalakshmi, K., Al Araimi, S., & Ali Saud Al Tobi, M. (2023). A review to do fishermen boat automation with artificial intelligence for sustainable fishing experience ensuring safety, security, navigation and sharing information for Omani fishermen. *Journal of Marine Science and Engineering, 11*(3), 630.

Fish Waste:
Coppola, D., Lauritano, C., Palma Esposito, F., Riccio, G., Rizzo, C., & de Pascale, D. (2021). Fish waste: From problem to valuable resource. *Marine Drugs, 19*(2), 116.

Figure Credits

Fig. 18.1   VisionDive on Shutterstock
Fig. 18.2   adailyodyssey on Shutterstock
Fig. 18.3   Left: Rob Jansen on Shutterstock; right: Nikki Gensert on Shutterstock
Fig. 18.4   Elise V on Shutterstock
Fig. 18.5   Damsea on Shutterstock
Fig. 18.6   Jukka Jantunen on Shutterstock
Fig. 18.7   Georgia Carini on Shutterstock
Fig. 18.8   Stephanie Rousseau on Shutterstock
Fig. 18.10 "Bycatch—tori lines (streamer lines)" by Michal Klajban is licensed under CC BY 4.0)
            Source: https://commons.wikimedia.org/wiki/File:Bycatch_-_tori_lines_(streamer_lines).svg
            Author: https://commons.wikimedia.org/wiki/User:Podzemnik
            License: https://creativecommons.org/licenses/by-sa/4.0/deed.en
Fig. 18.11 Dr. Erlijn van Genuchten
Fig. 18.12 Boris Pamikov on Shutterstock
Fig. 18.13 metamorworks on Shutterstock
Fig. 18.14 Magda Wygralak on Shutterstock
Fig. 18.15 Rosemarie Mosteller on Shutterstock
Fig. 18.16 Samjaw on Shutterstock
Fig. 18.17 Sinisa Botas on Shutterstock
Fig. 18.18 nobeastsofierce on Shutterstock
Fig. 18.19 Africa Studio on Shutterstock

Part IV
Conclusion

The earth is in bad condition; while solutions are put into place to address the three planetary crises and many individuals are taking action, still a lot more needs to be done to return to a healthy planet. This is because these crises are pushing the earth toward tipping points. Tipping points are points of no return because one environmental issue triggers further environmental issues, causing a cascade effect. For example, the melting of the Greenland Ice Sheet can trigger other issues at 1–3 °C (1.8–5.4 °F) temperature increase that push the temperature even higher above a catastrophic 4 °C (7.2 °F) increase.

Despite these urgent issues, one reason we are not doing more yet and at a faster rate is because of our psychology. In the Conclusion of A Guide to a Healthier Planet Volume 1, seven reasons for why our psychology is holding us back from implementing essential measures and changing our behavior are explained—and how they can be overcome. Another reason is our environmental mentality.

Chapter 19
Environmental Mentality

Abstract The Earth is facing three planetary crises, pushing it towards tipping points—points of no return. Despite efforts to address these crises, we are not doing enough because the 'grow now, clean up later' mentality is common. This mentality involves dealing with the negative consequences of today's behavior later. Cleaning up later is much more costly though and causes other issues such as higher health-care costs and reduced food production. That is why the opposite—an environmental protection mentality—is helpful in preventing environmental issues, health issues, resource scarcity, and counterproductive investments. It can be applied in all aspects of life, including at home and work, and turn our world's economy into a green economy.

Keywords Psychology · Mentality · Environmental psychology · Behavioral change · Environmental protection mentality · 'Grow now clean up later' mentality · Green economy · Sustainable investments · Health risks · Resource scarcity · Economic models · Innovation

Many feel that rapidly converting to an environmentally friendly economy is prohibitively expensive. The warnings that converting to an environmentally friendly economy is expensive impacts our environmental mentality. But these warnings are based on the predictions of standard economic models. Two of the most widely used models to quantify the costs and benefits of climate policy are DICE and RICE. These models look at the growth of a single new product or industry based on historic trends concerning the acceptance of new technologies. DICE stands for Dynamic Integrated model of Climate and the Economy; RICE stands for Regional Integrated model of Climate and the Economy. These models consider the money and hours of work required to produce a single

---

Credit: This chapter is based on the scientific article "The costs and benefits of environmental sustainability" by Paul Ekins and Dimitri Zenghelis. (Full citation is available at the end of the chapter).

product, without considering changes in other technologies that could influence product advancement and acceptance. As the scope of these models is limited, they fail to accurately predict trends in technologies that can be used to mitigate the three planetary crises.

Also, history doesn't provide an accurate model of our current technological changes. This is because we used to experience advances in one type of technology. But these days, technologies in multiple fields are advanced. This creates competition that drives further innovation. For example, in the energy sector, solar, wind, wave, fuel-cell, and geothermal technological developments compete to provide renewable energy. Traditional economic models, based on single advances in technology have proven inaccurate in predicting the growth of renewable energy industries: all world energy predictions since 1995 have significantly underestimated the growth of renewable energy and failed to predict dramatic price decreases. Prices decreased for instance for solar panels by 83% and for wind turbines by 35% since 2010.

While these models have a limited scope and therefore are not able to make accurate predictions, these models also suggest that growing economies now and cleaning the results later is the most suitable strategy (see Fig. 19.1). Unfortunately, the costs of this so-called 'grow now, clean up later' mentality are a lot higher than when implementing an environmental protection mentality.

Fig. 19.1 The 'grow now, clean up later' mentality causes a lot of costs, including financial costs

19.1 'Grow Now, Clean Up Later' Mentality

The 'grow now, clean up later' mentality seems attractive to some because they believe that costs for mitigation are higher than for solving issues later. But shockingly, the opposite is true. For example, an additional 0.5 °C temperature rise (0.9 °F) from 1.5 to 2 °C (2.7 to 3.6 °F) by 2100 is expected to cost between \$15 trillion and \$38.5 trillion, caused by a wide range of consequences. These are the worst costs of the 'grow now, clean up later' mentality:

19.1.1 Environmental Issues

The first cost of a 'grow now, clean up later' mentality is the cost of having to solve environmental issues. Dealing with environmental issues is often more expensive than preventing these issues. For example, dealing with environmental issues caused by littered waste is often more expensive than collecting and processing waste before it is distributed in the environment. And as some types of pollution accumulate over several years in the environment, the costs accumulate over the years as well.

For example, India applied the 'grow now, clean up later' concept and is an excellent example of the cost of trying to clean up later (see Fig. 19.2). Today, India spends about 5.7% of its Gross Domestic Product (GDP) on the resulting issues.

Fig. 19.2 Pollution India now has to deal with

The GDP is an economic indicator that measures the total value of all the goods created and services provided by a country in a year. Their GDP is $3.2 trillion, so 5.7% is $181 billion. Those $181 billion include costs for dealing with air pollution, inadequate water supply, sanitation and hygiene issues, and degraded farmlands, pasturelands, and forests. And costs for India are not only direct by paying to solve the issues, they also cause higher healthcare costs and reduced food production.

19.1.2 Counterproductive Investments

The second cost of a 'grow now, clean up later' mentality is the cost of having to make counterproductive investments. Counterproductive investments are for example investments in repairing or retrofitting technologies that harm the environment so that they continue to have harmful effects. Retrofitting means that new technology is added to reduce the harmful effects. The initial costs of repairing or retrofitting may be less but the eventual costs of cleaning environmental damage can be much higher.

For example, cleaning a factory that causes air pollution involves changing the equipment that causes the pollution. When broken equipment is repaired, the same amount of air pollution is emitted; when equipment is retrofitted, the amount of air pollution may be reduced; when equipment is replaced, the amount of air pollution may be completely prevented. All options cause financial investments but the environmental costs are lower the less pollution remains.

Also, purchasing equipment that uses more environmentally friendly technology immediately may be less expensive than purchasing equipment that requires retrofitting or replacement. This is because new technologies allow skipping the pollution phase. This concept of skipping obsolete technology by upgrading directly to the most efficient, environmentally friendly version is called leapfrogging. Leapfrogging allows for example developing nations to advance their technologies directly to the level of developed nations, avoiding the costs of cleaning that are required after using obsolete technology.

For example, developing nations can leapfrog inefficient centralized electricity grids by immediately investing in infrastructures that allow distributed energy generation (see Fig. 19.3). This is helpful in the transition toward generating energy from renewable sources as opposed to burning fossil fuels.

19.1.3 Health Risks

The third cost of a 'grow now, clean up later' mentality is the cost of having to deal with pollution-related diseases, such as lung cancer. Diseases triggered by or worsened by pollution cause 16% of deaths worldwide, about 254 million lives in 2015.

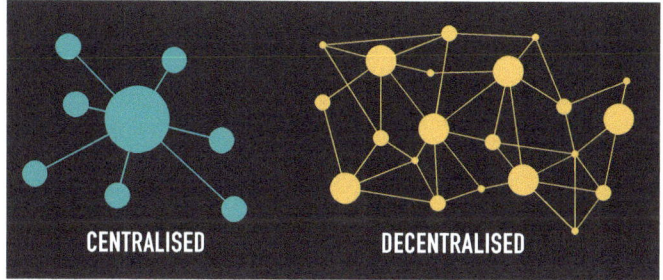

Fig. 19.3 Developing countries can leapfrog from centralized electricity systems based on fossil fuels to decentralized systems based on renewable energy

That makes pollution the leading cause of death, worse than tobacco, viruses and bacteria, or alcohol.

And while pollution creates costs in the number of deaths, it also creates financial costs. For example:

- the total welfare damage in 2015 is estimated at $4.6 trillion, about 6.2% of the global GDP
- water pollution costs $140 billion in lost wages annually
- water pollution causes $56 billion in healthcare costs annually

These costs in money, productivity, and health continue as long as the 'grow now, clean up later' mentality remains.

Luckily, these costs can be prevented as much of the pollution can be avoided using pollution control strategies. Many pollution control strategies have proven to be effective in developed countries and can be used in developing countries as well.

19.1.4   Lack of Resources

The fourth cost of a 'grow now, clean up later' mentality is the cost of having to deal with a lack of resources. Some resources are already lacking or scarce, other resources will inevitably run out sooner or later because they are non-renewable such as fossil fuels. This is because in general, our use of materials, including fossil fuels, has more than tripled since 1970, from 27 billion tons to 92 billion tons. This is about the same amount as about 4600 Great Pyramids of Giza in 1970 to 16,000 today! Also, Earth Overshoot Day was already on July 28th in 2022 and August 2nd in 2023. Earth Overshoot Day marks the day of the year when we have used more resources than can be replenished by nature.

Using more resources than can be replenished by nature also has indirect environmental costs. For example:

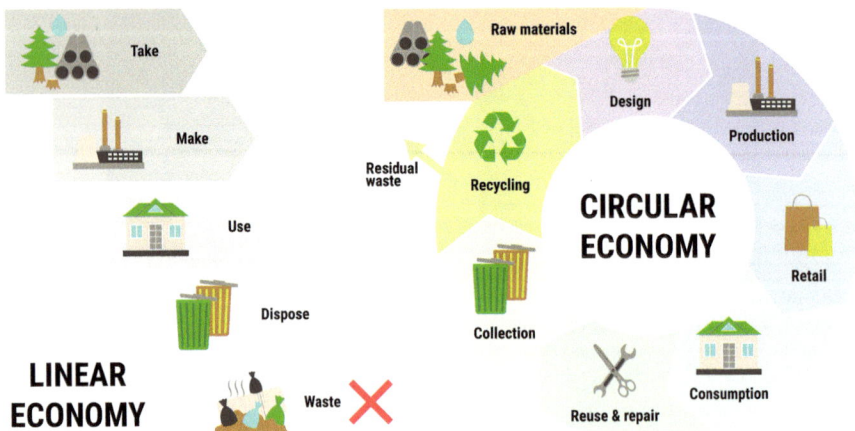

Fig. 19.4 Changing to a circular economy would reduce the need for additional resources by reusing existing resources

- agriculture and other biomass production is responsible for 90% of the loss of biodiversity
- agriculture uses more than 80% of fresh water available each year causing freshwater scarcity
- mining and extraction, including fossil fuels, cause pollution

When we continue these trends, our use of resources will double by 2060 to 190 billion tons (about 33,000 Great Pyramids of Giza); the amount of forest will decrease by 10% and other natural habitats by 20%. This means that, for example, pollution and clean freshwater scarcity will further increase and biodiversity will further decrease. These trends can be counteracted by converting our current linear economy to a circular economy. A linear economy means that resources are used and eventually go to waste. A circular economy means that no or hardly any resources are wasted and materials are used over and over again (see Fig. 19.4).

19.2   Environmental Protection Mentality

As the 'grow now, clean up later' mentality causes high costs, it is important to adopt the opposite mentality—an environmental protection mentality. An environmental protection mentality is a way of thinking that assigns the environment a high priority (see Fig. 19.5). This is important to prevent further environmental issues, health issues, resource scarcity, and counterproductive investments. Also, it allows us to prevent nature from taking care of our problems by depopulating and deindustrializing the planet.

An environmental protection mentality can be applied in all parts of our lives; at home, at work, etc. This is how this mentality benefits us:

Fig. 19.5 We need to change our world to a green economy

19.2.1 Changing Decisions and Behavior

The first benefit of an environmental protection mentality is that it changes our decisions and behavior. Decisions and behavior often only focus on how they benefit us. But when taking the environment into account as well, we are more likely to make decisions that improve the well-being of everyone, including nature. For example:

- an environmental protection mentality can inspire us to share more knowledge to stimulate further thoughts and innovations. That is for example why I did a 365 Sustainable Decisions Challenge and shared how I made my daily life more environmentally friendly: by sharing ideas from my daily life, others got inspired and adopted some of these changes in their daily life as well.
- an environmental protection mentality can inspire us to shift social norms. Social norms define acceptable group behavior. This is because an environmental protection mentality exposes ourselves and others to environmental concepts more often. This can trigger different decisions and behavior, which in turn contribute to changing public opinion and social norms. Such changes can be supported for instance by building cycle lanes and high-quality public transport, as they invite people to embrace new opportunities.
- an environmental protection mentality can inspire us to invest differently. Investments that once seemed safe to investors, pension fund managers, and insurance companies are becoming less valuable. For instance, investments in fossil fuels, including coal, become less valuable due to the transition to green energy sources.

19.2.2 Stimulating Innovation

The second benefit of an environmental protection mentality is that it stimulates and empowers innovation. Innovation is essential for our future ability to limit environmental degradation and resource use because it helps us replace environmentally unfriendly practices with more environmentally sound alternatives.

To empower innovation, an innovative idea needs to be followed up by research and development, a way of manufacturing or putting the idea into practice, and a network to distribute and sell the resulting product or service, where selling does not necessarily involve money. This means that innovation requires support, which is more likely provided by others who also have an environmental protection mentality. And when innovations are successful, they can trigger further investments in further ideas that can build on these positive experiences.

For example, knowledge from wind and solar power projects spreads more widely than knowledge from projects related to conventional energy sources. This further improves innovation in the energy sector but also in other industries such as the transportation industry.

19.2.3 Building New Industries and Technologies

The third benefit of an environmental protection mentality is that it stimulates the building of new industries and technologies (see Fig. 19.6). This is because having an environmental protection mentality makes it more likely that we are open to change.

Building new industries and technologies can be supported by governments by subsidizing research and development. They can also cover the initial costs of switching to new technologies through grants. This support is helpful for initiatives towards a healthier planet, as it allows focusing on creating change and initiating growth instead of being limited by challenges. Once those new industries and technologies hit a tipping point and are adopted, government support can shift to new initiatives. As a consequence, countries that invest early in environmentally friendly solutions have more success in future green markets.

And while new industries and technologies can be initiated from scratch, it is even more likely to be successful when a country invests in green technologies that are similar to their existing technologies. This is because this strategy allows taking advantage of existing factories and trained workers, using knowledge and technologies they already possess. As a consequence, an environmental protection mindset guides each country differently and encourages it to leverage the country's history and resources. A great practical example of the saying "standing on the shoulders of giants".

Fig. 19.6 An environmental protection mindset supports building new industries and technologies, such as wind turbines to generate electricity based on wind energy

Credit

This Chapter Is Based On:

Ekins, P., & Zenghelis, D. (2021). The costs and benefits of environmental sustainability. *Sustainability Science*, *16*, 949–965.

Figure Credits

Chapter 20
How We Can Take Action

Abstract As the 'grow now, clean up later' mentality is disastrous for our planet, it is important to combat this mentality and put the environmental protection mentality into practice. The good news is that many solutions are available and that many of these solutions can be put into practice by us as individuals in daily life. Even when these actions may seem small, they are essential, for example because they add up and inspire others to take action too

As the 'grow now, clean up later' mentality is disastrous for our planet, here are practical ideas of what you and I can do to combat the 'grow now, clean up later' mentality:

- Advocating for stronger pollution control laws at all levels of government
- Investing in clean technology whenever you are purchasing items that offer a choice
- Repairing broken objects
- Reusing materials as much as possible
- Bringing waste to recycling so that materials can be disposed of properly or—even better—reused

And as the environmental protection mentality is helpful in resolving the three planetary crises, here are practical ideas of what you and I can do to encourage an environmental protection mentality:

- Talking to others about environmental issues and solutions
- Encouraging new perceptions about environmental issues by expressing pro-environmental thoughts and concepts to friends and family
- Being a good example to others by looking for and applying ideas on how to make your daily life more environmentally friendly
- Buying environmentally friendly products and services instead of environmentally unfriendly alternatives, as this supports further development and acceptance of these products and services

Index